DAS TIERREICH

Eine Zusammenstellung und Kennzeichnung der rezenten Tierformen

The „TIERREICH" was initiated by the German Zoological Society in order to create a complete systematical treatment of all groups of the recent animal kingdom. After the World War II this undertaking was stopped because it seemed impossible to attain its completeness. In consequence of that, however, also ceased both incitement and possibility of publishing comprehensive revisions of large animal groups in a continued series. Therefore the Verlag Walter de Gruyter & Cie. (Berlin) came to the conclusion to revive the „TIERREICH". The original views will be maintained, the work, however, is planned now as an unconstrained sequence of systematical monographs without intending to treat all zoological categories according to a fixed plan. These monographs should not be restricted only to the belongs of taxionomy but also have regard to anatomical, biological, physiological, and other facts as far as they are able to characterize the group in question without breaking the frame of the whole work. All authors who are willing to publish systematical revisions of recent animal groups are invited to collaborate. According to the wishes of the authors, their contributions will be printed in English, French, or German. The authors will get an honorarium of DM 55,— for each 16 printed pages and 5 free copies after come out of their contributions. Further details may be gathered from the exemplaries no. 72 and subsequents published since 1958; by request the publishers will give any informations or send exact directions for the authors. Summary revisions of zoological groups (checklists) will although be published in the series „DAS TIERREICH".

Le „TIERREICH" fut fondé pa la Société Allemande de Zoologie dans le but de réaliser une présentation complète de la systématique de tous les groupes zoologiques récents. Cette entreprise fut suspendue après la deuxième guerre mondiale car la présentation complète que l'on s'était fixée pour but semblait impossible à réaliser. Mais, de ce fait, disparut, avec la possibilité, le stimulant nécessaire à la publication continue d'amples révisions de groupes zoologiques importants. C'est pourquoi, les éditions Walter de Gruyter & Co. (Berlin) ont pris la décision de faire renaître le „TIERREICH" sous l'optique originale pour ce qui est du fond, certes, mais uniquement sous la forme d'une série de monographies systématiques, sans lien étroit entre elles dont le but ne sera pas de traiter tous les groupes zoologiques d'après un plan préconçu. Ces monographies ne doivent pas simplement se borner au domaine de la taxionomie, mais tenir compte également des faits anatomiques, biologiques, physiologiques et autres dans la mesure où ils sont susceptibles de mieux caractériser le groupe zoologique en question sans pour cela déborder le cadré de l'ensemble de l'ouvrage. Sont invités à collaborer, tous les auteurs qui désireraient publier des révisions systématiques de groupes zoologiques récents. Conformément aux voeux des auteurs, leurs traités seront imprimés en anglais, en français et en allemand. Les honoraires des auteurs sont fixés à DM 55,— par placard de l'imprimerie (= 16 pages) et 5 exemplaires gratuits après parution de leurs traités. Pour plus amples détails, se reporter à les livraisons No. 72 et suivantes qui vient de paraître depuis 1958. Les éditeurs fournissent tous renseignements sur demande ou des indications précises à l'intention des auteurs. Dans la série ,,DAS TIERREICH", on publiera aussi des révisions sommaires de groupes zoologiques comme ,,Checklists".

Das ,,TIERREICH" wurde von der Deutschen Zoologischen Gesellschaft in der Absicht gegründet, eine vollständige systematische Darstellung aller rezenten Gruppen des Tierreichs zu schaffen. Nach dem zweiten Weltkrieg wurde das Unternehmen abgebrochen, da das Ziel der Vollständigkeit unerreichbar schien. Damit entfielen jedoch Anreiz und Möglichkeit, umfassende Revisionen größerer Tiergruppen in fortlaufenden Lieferungen zu veröffentlichen. Aus diesem Grunde hat sich der Verlag Walter de Gruyter & Co. (Berlin) entschlossen, das ,,TIERREICH" zwar grundsätzlich unter den ursprünglichen Gesichtspunkten, darüber hinaus aber lediglich als eine zwanglose Reihenfolge systematischer Monographien wieder aufleben zu lassen, ohne zu beabsichtigen, alle zoologischen Gruppen nach einem feststehenden Plan abzuhandeln. Diese Monographien sollen sich nicht ausschließlich auf die Belange der Taxionomie beschränken, sondern auch anatomische, biologische, physiologische und andere Tatsachen berücksichtigen, sofern sie geeignet sind, die betreffende Tiergruppe näher zu charakterisieren, ohne den Rahmen des Gesamtwerkes zu sprengen. Zur Mitarbeit sind alle Autoren eingeladen, die den Wunsch haben, systematische Revisionen rezenter Tiergruppen zu veröffentlichen. Entsprechend den Wünschen der Verfasser werden ihre Abhandlungen in Englisch, Französisch und Deutsch gedruckt. Die Autoren erhalten ein Honorar in Höhe von DM 55,— für jeden Druckbogen (= 16 Seiten) und 5 Sonderdrucke nach Erscheinen ihrer Beiträge. Weitere Einzelheiten mögen den seit 1958 erschienenen Lieferungen Nr. 72 und folgenden entnommen werden. Auf Wunsch erteilen die Herausgeber jegliche Auskünfte oder geben genaue Anweisungen an die Autoren ab. In der Reihe ,,DAS TIERREICH" erscheinen auch summarische Bearbeitungen von größeren Tiergruppen in Form sogenannter ,,Checklists".

Die Herausgeber

†Prof. Dr. Robert Mertens (Wirbeltiere)
Natur-Museum und Forschungs-Institut Senckenberg
Senckenberg-Anlage 25, 6000 Frankfurt am Main 1

Prof. Dr. Willi Hennig (Wirbellose)
Staatliches Museum für Naturkunde
Arsenalplatz 3, 7140 Ludwigsburg

Dr. Heinz Wermuth (Schriftleiter)
Staatliches Museum für Naturkunde
Arsenalplatz 3, 7140 Ludwigsburg

DAS TIERREICH

Eine Zusammenstellung und Kennzeichnung der rezenten Tierformen

Herausgeber

ROBERT MERTENS (Wirbeltiere)

WILLI HENNIG (Wirbellose)

Schriftleitung
HEINZ WERMUTH

Unter Mitarbeit zahlreicher Spezialisten

„Πάντα ῥεῖ“ „Sine systemate chaos“

Lieferung 99

Lepidoptera Noctuiformes

Agaristidae III

(American Genera)

Bearbeitet von

Prof. Dr. S. G. Kiriakoff, Gent

Mit 26 Figuren im Text

Walter de Gruyter · Berlin · New York
1977

Das Tierreich	Berlin	Lieferung 99	Seite I—VI, 1—86	1977

ISBN 3 11 006859 1

Printed in Germany.
Satz: IBM-Composer Walter de Gruyter & Co., Berlin.
Druck: Karl Gerike, Berlin.
Bindearbeiten: Dieter Mikolai, Berlin.

Lepidoptera Noctuiformes

Agaristidae III

(American Genera)

Bearbeitet von

Prof. Dr. S. G. Kiriakoff

Systematic Index

Family **A g a r i s t i d a e**

III. American Genera

Parothria 2
P. ecuadoriana 3

Rhosus 4
R. posticus 5
R. albiceps 6
R. leuconoë 6
R. pulverosus 7
R. colombianus 7
R. aguirrei 7
R. ovatus 8
R. stornianus 8
R. denieri 8
R. spadiceus 8
R. isabella 9

Cicaucula 9
C. peruviana 10

Erocha 11
E. mummia 12
E. leucodisca 13
E. leucotelus 13
E. semiviridis 14
E. elaeina 15
E. picta 15
E. affinis 15
E. irrorata 16
E. dolens 16
E. trita 17

Aucula
A. josioides 18
A. buprasia 19
A. hilzingeri 19
A. magnifica 20
A. janeira 20
A. sublata 21
A. schausi 21

Gerrodes 22
G. longipes 22
G. minatea 24

Xerociris 24
X. wilsonii 25

Orthia 26
O. augias 27

Seirocastnia 28
S. amalthea 29
S. panamensis 29
S. volupia 30
S. tribuna 30
S. nocturna 31
S. meridiana 31
S. columbina 32
S. extensa 32
S. albifascia 32

Epithisanota 32
E. sanctijohannis 33

Darceta 35
D. proserpina 36
D. primulina 37

D. ophideres 37
D. grandimacula 37
D. falcata 38
D. haenschi 39
D. severa 39

Copidryas 40
C. gloveri 41
C. cosyra 42

Euscirrhopterus 42
E. poeyi 43
E. klagesi 45
E. walkeri 45
E. w. walkeri 46
E. w. tucumanus 46
E. disciferus 46

Schalifrontia 47
S. furcifer 47

Alypia 48
A. octomaculata 49
A. wittefeldii 51
A. matuta 51
A. dipsaci 52
A. langtoni 52
A. ridingsi 52
A. mariposa 53
A. australis 53
A. ornata 54
A. pampeana 54
A. lulesa 54
A. maccullochii 54
A. m. maccullochii 55
A. m. lorquini 55
A. m. similis 55
A. disparata 56
A. brannani 56

Alypioides 56
A. bimaculata 57
A. radians 58
A. walkeri 59
A. geronimo 59

Psychomorpha 59
P. epimenis 60
P. euryrhoda 61

Phasidia 61
P. contraria 62

Neotuerta 63
N. sabulosa 63
N. s. sabulosa 65
N. s. collectiora 65
N. platensis 65
N. p. platensis 65
N. p. bolivar 66
N. hemicycla 66
N. lycaon 66

Melanchroiopsis 67
M. acroleuca 67
M. noctilux 68
M. mardava 69

Caularis 69
C. undulans 70
C. lunata 71
C. jamaicensis 71

Gerra 72
G. radicalis 73
G. sevorsa 74
G. brephos 74
G. aelia 74
G. pulchra 75
G. lunata 75
G. sophocles 75

Parathisanotia 76
P. grata 76

Euthisanotia 77
E. unio 78
E. brevipennis 80

Pseudalypia 80
P. crotchii 81

Letaba 81
L. noa 81

Eupseudomorpha 82
E. brillians 82

Introduction

The author deeply regrets his inability to keep this third and last part of his work on Agaristidae to the standard of the previous parts. It happens that E. L. TODD of the U.S. Department of Agriculture is currently working on a revision of the genera *Aucula*, *Erocha* and (with J. G. FRANCLEMONT) *Darceta*, and besides is planning to describe a few new species. The results apparently are not to be expected before some long time. Unfortunately, the author's time being strictly limited, he was unable to postpone any longer the preparation of the present part as he already had postponed it for a year to comply with the wish expressed by Mr. TODD. As a result, the genera named above have been treated in the classical way although it results both from the author's own research and from the correspondence with Mr. TODD that they are composite and contain some non congeneric species.

On the other hand, the author was unable to procure (notwithstanding repeated queries) slides or photographs of the genitalia of the type specimens of the following, viz. *Pseudalypia crotchii*, *Letaba noa* and *Eupseudomorpha brillians*, those being preserved in American museums and not represented in Europe. In consequence of which, the three genera are considered in the present work as *incertae sedis.*

Fam. Agaristidae

American Genera

Key to the genera

1 Areole present 2
1′ Areole absent . . . . ***Darceta (partim)***, p. 35, ***Parothria***, p. 2
2 Fore wings with vein 10 arising from the extremity of areole . . ***Orthia (sometimes)*** p. 26
2′ Fore wings with vein 10 branching off from the upperside of areole 6
3 Fore wings with vein 6 branching off from areole 4
3′ Fore wings with vein 6 arising from below the upper cell-angle 5
4 Antennae strongly dilated towards the extremity, without pectinations . . . . ***Euscirrhopterus***, p. 42
4′ Antennae bipectinate to the extremity . . . . ***Xerociris***, p. 24
5 Abdomen dorsally clothed with rough hair . . . . ***Orthia***, p. 26
5′ Abdomen dorsally smoothly scaled . . . . ***Phasidia***, p. 61
6 Eyes hairy . . . . ***Schalifrontia***, p. 47
6′ Eyes smooth 7
7 Frons with flattened process bearing three horns 8
7′ Frons without such process, generally with conical truncate prominence with raised rim at the extremity 9

8 Abdominal crests on basal segments; tibiae fringed with long hair . . . . . ***Copidryas*, p.** 40
8′ Abdomen with large crest on urite I only; tibiae with large tufts and long hairfringe . . ***Cisaucula*, p.** 9
9 Abdomen with crests on basal segments **10**
9′ Abdomen without crests, or with crest on urite I only **13**
10 Abdomen with crests on 4–5 urites . . . . ***Gerra*, p.** 72
10′ Abdomen with crests on 2–3 basal urites **11**
11 Antennae strongly dilated towards the extremity . . . . ***Neotuerta*, p.** 63
11′ Antennae ciliated or simple **12**
12 Antennae ciliated . . . . ***Euthisanotia*, p.** 77
12′ Antennae simple . . . . ***Parathisanotia*** p. 76 and ***Epithisanotia****) p. 32
13 Frons with trilobate prominence, the median part the longest . . . ***Eupseudomorpha*, p.** 82
13′ Frons with prominence differently shaped **14**
14 Frons with truncate conical prominence, tip flat with scarcely raised rim . . ***Letaba*, p.** 81
14′ Frons with prominence differently shaped **15**
15 Antennae bipectinated, at least in ♂ **16**
15′ Antennae differently shaped **20**
16 Abdomen without crests **17**
16′ Abdomen with basal crest **19**
17 Frons with large conical pointed prominence . . . . ***Psychomorpha*, p.** 59
17′ Frons with prominence differently shaped **18**
18 Hind wings with veins 6 and 7 short stalked . . . . ***Rhosus*, p.** 4
18′ Hind wings with veins 6 and 7 from a point . . . . ***Aucula*, p.** 17
19 Frons with rounded prominence; at centre a truncate conical process with raised rim and a corneous plate below . . . . ***Erocha*, p.** 11
19′ Frons with large truncate conical prominence, with a small apical process with raised rim . . . . ***Caularis*, p.** 69
20 Antennae ciliated or laminated . . . . ***Gerrodes*, p.** 22
20′ Antennae simple, more or less dilated towards the extremity **21**
21 Antennae distinctly dilated towards the extremity . . . . ***Alypia*, p.** 48
21′ Antennae slightly dilated towards the extremity **22**
22 Frons with triangular corneous process notched at sides . . . . ***Pseudalypia*, p.** 80
22′ Frons with the prominence differently shaped **23**
23 Frons with a hairtuft and a small truncated prominence with slightly raised rim . . . . ***Seirocastnia*, p.** 28
23′ Frons without hairtuft, and with the prominence differently shaped **24**
24 Fore wings slightly falcate . . . . ***Darceta (partim)*, p.** 35
24′ Fore wings not falcate **25**
25 Abdomen without crests . . . . ***Alypioides*, p.** 56
25′ Abdomen with large basal crest . . . . ***Melanchroiopsis*, p.** 67

*) These two genera differ only in the structure of ♂ genitalia.

Genus **Parothria** HAMPSON

1901 *Parothria* HAMPSON, Cat. Lepidopt. Phal. Brit. Mus., **3**:654–655. – Species typica: *Othria ecuadorina* WESTWOOD (by original designation).

Diagnosis: Proboscis fully developed; antennae distinctly dilated towards the extremity; palpi upturned, not reaching vertex, fringed with long hair in front; the 3rd joint short and clothed with hair; frons with a small rounded truncate conical prominence, with raised rim at the extremity; hind tibiae fringed with long hair. Nervation: in fore wings, vein 2 from about five-sevenths of cell; 3 and 4 approximate; distance between 4 and 5 about twice as long as that between 3 and 4; discocellulars oblique, faintly incurved; 6 from just below the upper cell-angle; areole absent; 7, 10, 8 + 9 stalked. In hind wings, vein 2 from slightly beyond three-quarters of cell; 3, 4 and 6, 7 from the cell-angles; discocellulars incurved; 5 obsolescent, from middle of discocellulars.

Male genitalia: Uncus about three-fifths of costa in length, compressed, well arched basally, then very faintly arched, with the tip acute, but not forming a true hook. Vinculum moderately broad, with the fold as broad as the main part and bearing laterally a few hairs. Valva very broad; costa nearly straight, faintly outcurved only just beyond middle; apex broadly rounded; sacculus broadly outcurved; fold of sacculus broad basally, but strongly narrowing distally and ending in a short, wedge-shaped harpe. Aedeagus slightly shorter than sacculus, rather robust, boomerang-shaped, with the proximal three-fifths spoon-shaped; fultura inferior with the proximal margin produced and excised at middle, narrowing distally, with the sides slightly incurved; distal angle slightly produced. Saccus short semi-oval.

Distribution: Ecuador.

Parothria ecuadorina (WESTWOOD)

(fig. 1)

1877 *Othria ecuadorina* WESTWOOD, Trans. Linn. Soc., Zool. (2):202. – Terra typica: Ecuador (Type, ♂, in Oxford Museum).
1901 *Parothria ecuadorina* – HAMPSON, Cat. Lepidopt. Phal. Brit. Mus. 3:655, fig. 289.

Diagnosis: Body and wings black; some yellow hair at base of fore and mid legs; hind femora fringed with yellow hair. Fore wings with a narrow oblique orange band from below costa just beyond cell to above tornus, slightly forked at

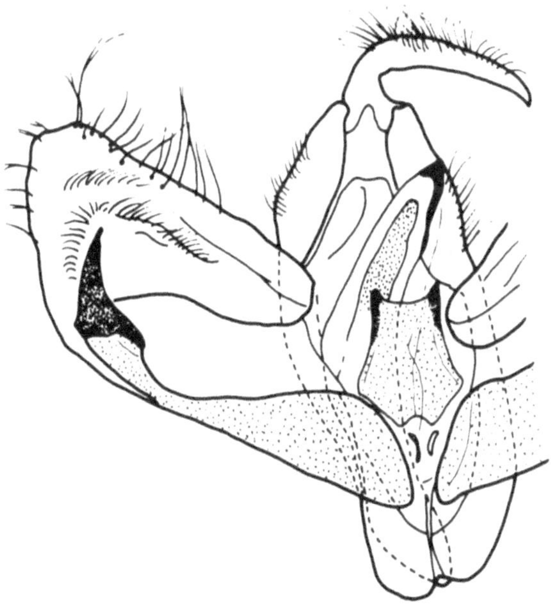

Fig. 1. *Parothria ecuadorina* (WESTWOOD). 8 Leiden Mus.

the lower extremity. Hind wings with a wedge-shaped orange patch from base extending from the subcostal vein to the submedian fold and to well beyond cell, its distal edge produced to an angle between veins 2 and 3. Length of fore wing 23 mm.

Distribution: Ecuador.

Genus **Rhosus** WALKER

1854 *Rhosus* WALKER, List Lepidopt. Het. Brit. Mus., **2**:359. – Species typica: *Rhosus posticus* WALKER.
1855 *Mystrocephala* HERRICH-SCHÄFFER, Außereurop. Schmetterl.,:13; fig. 435. – Species typica: *Mystrocephala quadrimacula* = *Rhosus posticus* WALKER.
1874 *Pycnodontis* FELDER, Reise Novara, :8. – Species typica: *Pycnodontis spadicea* FELDER.

Diagnosis: Proboscis fully developed; antennae slightly dilated beyond the middle, with short pectinations in ♂, almost simple in ♀; palpi bent downwards, with short pilosity; frons bearing a truncate conical prominence with a raised ring at the extremity; head and thorax clothed with rough hair and scales; tibiae fringed above with rough hair; abdomen smoothly scaled, without crests. Nervation: In fore wings, vein 2 from about three-quarters of cell; 3, 4, 5 approximate, equidistant; discocellulars well incurved; areole narrow; 6 from the upper cell-angle; 7 and the short stalk of 8 + 9 from the extremity of the areole; 10 from the upper angle thereof; costa straight except at the base. In hind wings, vein 2 from two thirds of cell; 3, 4 from the lower cell-angle; discocellulars oblique, faintly incurved; 5 obsolescent, from middle of discocellulars; 6, 7 short-stalked.

Male genitalia: Uncus about three-quarters of costa in length, compressed, faintly arched, ending in a blunt point. Tubus analis short but broad, with only ventral sclerification. Vinculum very broad, but with the fold narrow and occupying only about one-third of its length, just beyond the middle; pilosity rather sparse. Valva very broad, upturned; costa nearly straight; apex rounded; sacculus regularly outcurved; its fold narrow, reaching slightly beyond two fifths of the valva, ending in a robust, nearly straight harpe.Aedeagus about one-quarter longer than the costa, club-shaped, but with the proximal extremity spoon-shaped; a subterminal patch of minute cornuti; a terminal small horn-shaped process; fultura inferior with its proximal margin outcurved, especially at middle; sides parallel; distal margin slightly bilobate; fultura superior broad, with the distal angles slightly produced and rounded. Saccus hardly developed.

Distribution: Central and South America

Key to the species

1 Hind wings yellowish with a black terminal band 2
1′ Hind wings uniform black, or with a yellowish patch 3
2 Fore wings with the postmedial band intersected by black nervural streaks . . . . *R. leuconoë*, p. 6
2′ Fore wings with the postmedial band not intersected with black nervural streaks . . . . *R. pulverosus*, p. 7
3 Hind wings with a yellowish postmedial patch, sometimes only on the underside in ♂ . 4
3′ Hind wings without pale postmedian patch 6
4 Hind wings with a white or yellowish patch on both sides in both sexes 5

4′ Hind wings with a yellowish patch sprinkled with black, only on the underside in ♂, on both sides in ♀ . . . . *R. colombianus*, p. 7
5 Occiput, collar and tegulae deep red-brown, sprinkled with yellowish-white . . . . *R. posticus*, p. 5
5′ Occiput, collar and tegulae white . . . . *R. albiceps*, p. 6
6 Fore wings with a yellowish medial spot in interspace I . . . . *R. aguirrei*, p. 7
6′ Fore wings without pale medial spot in interspace I 7
7 Hind wings with two or three pale streaks on the underside 8
7′ Hind wings without pale streaks on the underside 10
8 First joint of palpi, and fore and median coxae red . . . . *R. stornianus*, p. 8
8′ Palpi and legs without red 9
9 Palpi and legs black; fore wings black with grey reflections; underside of hind wings with whitish streaks on the media and cubitalis . . . . *R. denieri*, p. 8
9′ Palpi black; coxae with yellowish pilosity; fore wings deep chocolate brown sprinkled with yellowish; underside of hind wings with whitish streaks on the costa, media and vein 1 . . . . *R. ovatus*, p. 8
10 Wings with the ground colour deep red-brown . . . . *R. spadiceus*, p. 8
10′ Wings with the ground colour brick-red, more or less sprinkled with ochreous . . . . . . *R. isabella*, p. 9

Rhosus posticus WALKER

(fig. 2)

1854 *Rhosus posticus* WALKER, List. Lepidopt. Het. Brit. Mus., **2**:360. – Terra typica: Brazil, Rio de Janeiro (Type, ♂, in British Museum).
1854 *Melanchroia clavigera* WALKER, List Lepidopt. Het. Brit. Mus., **2**:390. – Terra typica: Brazil, Rio de Janeiro (Type in British Museum).

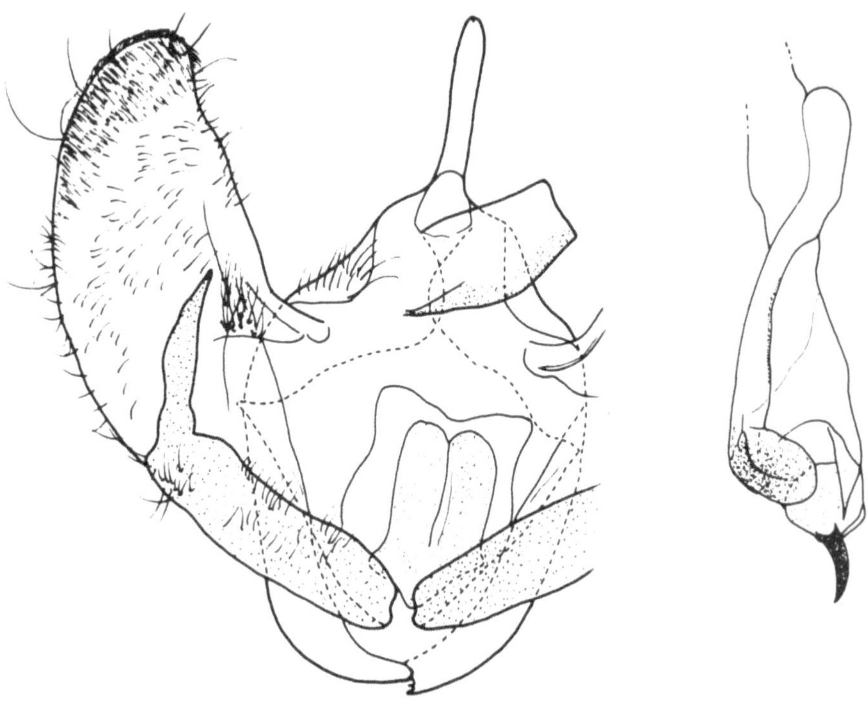

Fig. 2. *Rhosus posticus* WALKER. BM Noct. 8412. Type

1854 *Melanchroia aeruginosa* WALKER, List Lepidopt. Het. Brit. Mus., 2:391. – Terra typica: Brazil, Rio de Janeiro (Type in British Museum).
1855 *Mystrocephala quadrimacula* – HERRICH-SCHÄFFER, Außereurop. Schmetterl., fig.: 435.
1910 *Rhosus postica* – HAMPSON, Cat. Lepidopt. Phal. Brit. Mus., 9:412, fig. 195.
1912 *Mystrocephala postica* – STRAND, in Lepidopt. Cat., 5:50.

Diagnosis: Body and wings deep red-brown sprinkled with ochreous-white scales; proboscis yellow; coxae ochreous white in ♂; abdomen with an ochreous-white lateral line and ventral bands in ♂; anal tuft orange. Fore wings with an elliptical oblique ochreous white spot beyond the cell between veins 3 and 7; cilia tipped with white at the apex. Hind wings with a wedge-shaped ochreous-white patch from middle of cell to beyond the cell-tip; more or less developed ochreous-white streaks in the submedian fold, and on the anal area from near base to beyond the middle, both absent in ♀; cilia tipped with at the apex and towards the tornus in ♀. Length of fore wing 20 mm.

Distribution: Brazil; possibly Argentina.

Rhosus albiceps (DRAUDT)

1919 *Mystrocephala albiceps* DRAUDT in SEITZ, Großschmetterl. Erde, 11:10; pl. 5b. – Terra typica: Colombia, Muzo. (Type in Berlin Museum).

Diagnosis: Similar to *R. posticus*, differing in occiput, collar and tegulae being white; the yellow pilosity of the legs very long; sides of the second and third abdominal segments and the underside of the abdominal tuft orange. Hind wings with only a narrow, wedge-shaped ochreous white spot. Length of fore wing 17 mm.

Distribution: Colombia.

Rhosus leuconoë (FELDER)

1868/74 *Pycnodontis leuconoë* FELDER, Reise Novara: 107, fig. 13. – Terra typica: Colombia, Bogotà (Type not designated).
1910 *Rhosus leuconoë* – HAMPSON, Cat. Lepidopt. Phal. Brit. Mus., 9:412.
1912 *Mystrocephala leuconoë* – STRAND, in Lepidopt. Cat., 5:50.

Diagnosis: Body black sprinkled with white; proboscis yellow; abdomen with whitish lateral stripes and slight orange streaks towards the extremity; the latter orange; underside suffused with whitish. Fore wings deep chocolate-brown sprinkled with ochreous-white; two divergent white streaks in the cell, not reaching the discocellulars; an obliquely curved band beyond the cell between veins 2 and 7, divided into spots by veins. Hind wings ochreous tinged with orange; veins black; a terminal black band expanding on the apical area. Length of fore wing 28 mm.

Distribution: Colombia.

Rhosus pulverosus (ROTHSCHILD)

1854 *Rhosus posticus* WALKER (partim), List Lepidopt. Het. Brit. Mus., **2**:360.
1896 *Pycnodontis pulverosa* ROTHSCHILD, Novit. Zool., **3**:98, pl. 15, fig. 34. – Terra typica: Panamà (Type in British Museum).
1910 *Rhosus pulverosa* – HAMPSON, Cat. Lepidopt. Phal. Brit. Mus., **9**:412.
1912 *Mystrocephala pulverosa* – STRAND, in Lepidopt. Cat., **5**:50.

Diagnosis: Body and wings black; body sprinkled with ochreous-white; proboscis yellow; terminal segment of the abdomen and anal tuft orange. Fore wings very strongly sprinkled with ochreous; an oblique elliptical ochreous-white spot beyond the cell between veins 2 and 7, with traces of a line beyond it angled at vein 4 and oblique towards the costa and tornus. Hind wings with a large ochreous-white patch extending from near the base to beyond the middle of vein 3, and from the subcostal vein to the dorsum, with dark veins. Length of fore wing 22 mm.

Distribution: Panamà, Venezuela.

Rhosus colombianus HAMPSON

1910 *Rhosus colombiana* HAMPSON, Cat. Lepidopt. Phal. Brit. Mus., **9**:413; pl. 146, fig. 1. – Terra typica: Colombia (Type, ♀, in British Museum, Type, ♂, in Berlin Museum).
1912 *Mystrocephala colombiana* – STRAND, in Lepidopt. Cat., **5**:50.

Diagnosis: Head and thorax deep chocolate-red sprinkled with yellowish-white; proboscis yellow; abdomen black with slight dorsal yellowish sprinkling; extremity orange; underside with pale yellow bands in male. Fore wings chocolate-red sparsely sprinkled with pale yellow; an oblique yellowish-white band beyond the cell between veins 2 and 7. Hind wings blackish-brown; ♀ with a yellowish patch sprinkled with black, just beyond the cell, running obliquely from vein 3 to vein 6; cilia yellowish at the extremities. Length of fore wing 18–20 mm.

Distribution: Colombia.

Rhosus aguirrei (BERG)

1882 *Alypia aguirrei* BERG, Ann. Soc. Argent., **13**:174. – Terra typica: Argentina, Buenos Aires (Type in Museo Buenos Aires).
1910 *Rhosus aguirrei* – HAMPSON, Cat. Lepidopt. Phal. Brit. Mus., **9**:413; pl. 146, fig. 2.
1912 *Mystrocephala aguirrei* – STRAND, in Lepidopt. Cat. **5**:50.

Diagnosis: Body and fore wings blackish-brown sprinkled with greyish-white; anal tuft of the abdomen orange. Fore wings with red streaks on the base of the costa, in the cell, and above the submedian fold, the last extending from the base to the terminal band, which is somewhat maculate, expanding slightly at the apex, angled inwards at vein 5, and ending at the submedian fold; a yellowish medial spot above vein 1; a large spot beyond the cell from vein 8 to below vein 3, somewhat oblique and constricted at the middle. Hind wings dark chocolate brown. Length of fore wing 17 mm.

Distribution: Argentina.

Rhosus ovatus (ROTHSCHILD)

1896 *Pycnodontis ovata* ROTHSCHILD, Novit. Zool., 3:98; pl. 15, fig. 33. – Terra typica: Brazil, Sao Paulo (Type in British Museum).
1896 *Rhosus unipuncta* SCHAUS, J. New York entomol. Soc., 4:153. – Terra typica: Brazil, Sao Paulo (Type in U.S. National Museum).
1910 *Rhosus ovata* – HAMPSON, Cat. Lepidopt. Phal. Brit. Mus., 9:414.
1912 *Mystrocephala ovata* – STRAND, in Lepidopt. Cat., 5:50.

Diagnosis: Body black, sprinkled with yellowish-white; proboscis orange; coxae and hind tibiae with yellowish pilosity, tarsi ringed with white; abdomen with the extremity orange; underside with pale yellow bands. Fore wings deep chocolate-brown, sprinkled with yellowish-white; an oblique elliptical yellowish-white spot beyond the cell between veins 3 and 8; cilia slightly white at the apex. Hind wings blackish brown; cilia white at the extremities. Length of fore wing 15–19 mm.

Distribution: Brazil (Bahia, Sao Paulo); Argentina.

Rhosus stornianus (KÖHLER)

1936 *Mystrocephala storniana* KÖHLER, Physis, 12:72; pl. upper right figure. – Terra typica: Argentina, Concordia (Type in Museum Buenos Aires).

Diagnosis: Body and wings bluish-black; anal hair brick-red; the first joint of palpi, and fore and mid-coxae brick-red with a crimson tinge. Fore wings with a squarish white spot with pearly gloss. Hind wings unspotted; cilia white. Nearest to *Rh. ovatus* (Rothschild).

Distribution: Argentina.

Rhosus denieri (KÖHLER)

1936 *Mystrocephala denieri* KÖHLER, Physis, 12: 72–73; pl. upper left figure. – Terra typica: Argentina, Rio Porongal, Salto. (type in Museum Buenos Aires).

Diagnosis: Nearest to *R. stornianus.* Body and wings brownish-black; a white subocular spotlet; palpi and legs black; anal hair orange. Fore wings with a greyish gloss; a white band just beyond the middle, not reaching the costa nor the tornus, but encroaching into the discocellulars; cilia white at the apex. Hind wings with blue reflections; cilia white to near the anal angle. Length of fore wing 18 mm.

Distribution: Argentina.

Rhosus spadiceus (FELDER)

1868/74 *Pycnodontis spadicea* FELDER, Reise Novara: pl. 107, fig. 12. – Terra typica: Colombia, Bogotà (Type not designated).
1883 *Alypia spadicea* – BERG, Ann. Soc. Argent., 15:153.
1910 *Rhosus spadicea* – HAMPSON, Cat. Lepidopt. Phal. Brit. Mus., 9:414.
1912 *Mystrocephala spadicea* – STRAND, in Lepidopt. Cat., 5:50.

D i a g n o s i s : Body and wings deep red-brown, slightly sprinkled with ochreous-white; proboscis yellow; throat, coxae in male, and ventral bands on the abdomen in male, ochreous-white; anal hair orange. Fore wings with an oblique ochreous-yellow band beyond the cell from below the costa to vein 2 near the termen. Hind wings with cilia yellow apically. Length of fore wing 18–20 mm.

D i s t r i b u t i o n : Colombia, Bolivia, Ecuador, Perù.

Rhosus isabella DOGNIN

1898 *Rhosus isabella* DOGNIN, Ann. Soc. entomol. Belg., **42**:345. – T e r r a t y p i c a : Ecuador (Type in U.S. National Museum).
1912 *Mystrocephala isabella* – STRAND, in Lepidopt. Cat., **5**:50.

D i a g n o s i s : Head black, sprinkled with white; thorax deep red brown, sprinkled with ochreous; abdomen black, sprinkled with ochreous, and with a lateral series of white points; extremity orange. Fore and hind wings brick-red; fore wings thickly sprinkled with ochreous; a very oblique ochreous band beyond the cell from vein 7 to below vein 3; cilia black. Hind wings slightly sprinkled with ochreous; cilia ochreous. Length of fore wing 25 mm.

D i s t r i b u t i o n : Ecuador.

Genus **Cisaucula** TODD

1966 *Cisaucula* TODD, Proc. U.S.Nat. Mus., **120**:14–15. – S p e c i e s t y p i c a : *Copidryas peruviana* DRUCE (by original designation).

D i a g n o s i s : Proboscis fully developed; antennae filiform; palpi short, slightly oblique, with the second and the third segments fringed ventrally with long hair, the third segment slightly hanging, clothed with appressed scales; thorax clothed with hair and hairlike scales mixed; no thoracic crests; legs with large tufts and fringes of long white hair; abdomen with large dorsal crest on the first urite. Nervation: in fore wings, vein 2 arising from two-thirds of cell; 3, 4 and 5 separated, 5 slightly more distant from 4 than the latter form 3; 6 from the lower cell-angle; areole present; 7 short stalked with the short stalk of 8 + 9, from the extremity of the areole; 10 from the upper margin thereof. In hind wings, vein 2 arising beyond two-thirds of cell; 3 and 4 from the lower cell-angle; 5 obsolescent, from middle of the discocellulars; 6 and 7 short stalked.

Male genitalia: Uncus short, about one-third of costa in length, slightly contricted after the base, then broadening, rounded terminally, with a slight terminal hook. Tubus analis broad. Vinculum narrow; fold rather broad, with short sparse pilosity. Valva rather broad, slightly narrowing terminally; apex rounded; fold of sacculus relatively narrow, about one-third of valva at the base, but somewhat broadening at two-thirds, reaching two-thirds of valva, and ending in a very long, rather robust harpe, bent at right angle at middle, and reaching beyond the apex of the valva. Aedeagus about as long as onehalf of costa; basal half broadened and flattened;

terminal half slender, ending in a small knoblet, and covered for the most part with minute cornuti; fultura inferior forming a broad parallelogram with rounded angles. Saccus very short, broadly rounded.

Distribution: Peru, Bolivia.

Cisaucula peruviana (DRUCE)

(fig. 3)

1910 *Copidryas peruviana* DRUCE, Ann. Mag. Nat. Hist., (8) **6**:168. – Terra typica: Perù, Santo Domingo, 6000 ft. (Lectotype, ♂, in British Museum).
1966 *Cisaucula peruviana* – TODD, Proc. U.S.Nat. Mus., **120**:15; fig. 21.

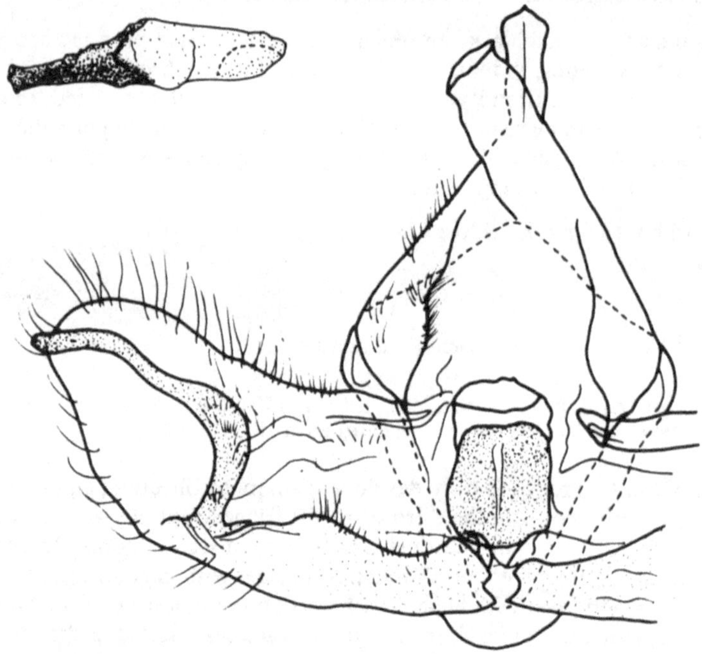

Fig. 3. *Cisaucula peruviana* (DRUCE). BM Noct. 8495. Type

Diagnosis: Head, collar, tegulae and thorax brown mixed with grey hair; antennae and palpi black; legs yellow with a fringe of long white hair; abdomen yellow with the basal crest black; underside greyish yellow. Fore wings pale brown, thickly sprinkled with white and greenish; dorsal area brown from base to tormus; a small brown cell-spot; two postmedial zigzag black lines from the costa to the dorsum; cilia chequered brown and white. Hind wings chrome-yellow, with a black marginal band from the apex to near the anal angle; cilia chequered black and grey. Female rather darker. Length of fore wing 20 mm.

Distribution: Perù, Bolivia.

Genus **Erocha** WALKER

1854 *Erocha* WALKER, List Lepidopt. Het. Brit. Mus., **2**:318. – Species typica: *Bombyx mummia* CRAMER (by monotypy).
1875 *Leiosoma* FELDER, Reise Novara,: 6. – Species typica: ***Euthisanotia leucotelus*** WALKER (by monotypy).
1896 *Leonides* DRUCE, Biol. Centr. Amer., Heterocera **2**:328. – Species typica: *Euthisanotia leucotelus* WALKER (by monotypy).

Diagnosis: Proboscis fully developed; antennae of ♂ bipectinate, with moderately long branches and serrate apex, antennae of ♀ serrate (in *E. dolens* Druce bipectinate to apex); palpi porrect, reaching to just beyond the frons; the second segment moderately fringed with hair, the third short; frons with a rounded prominence with a truncate conical process at the centre, with a raised rim and a corneous plate below it; eyes large, round. Thorax without crests, clothed with hair and scales mixed; tibiae moderately fringed with hair. Abdomen with a dorsal crest at base only. Venation: in fore wings, vein 2 arising from about three-quarters of cell; 3, 4, 5 separated, equidistant; discocellulars faintly incurved; 6 from the upper cell-angle; areole small; 7 and the short stalk of 8 + 9 from its extremity; 10 from its upper margin, near the extremity. In hind wings, vein 2 arising from about two-thirds of cell, 3, 4, and 6, 7 from the cell-angles (in *E. dolens* 6, 7 short stalked), discocellulars incurved; 5 obsolescent, from middle thereof; 8 anastomosing for a short while with cell.

Male genitalia (*Erocha mummia*): Uncus less than one-half of costa in length, compressed, arched only basally, ending in a point; pilosity dorsal, very dense and long. Tubus analis broad, with only dorsal sclerification. Valva broad, upcurved, with costa and sacculus nearly parallel, both slightly outcurved beyond two-thirds of valva; costa (sometimes) bearing a hornlike subterminal process; apex rounded; fold of sacculus broad basally, narrowing distally, reaching to middle of valva and bearing terminally a small triangular harpe; valvula bearing at middle a small, short triangular process. Aedeagus relatively very short, less than half of costa, robust, nearly straight, bearing terminally a slender, arched process, typically about four-fifths of costa in length; fultura inferior semi-circular, with the proximal edge rounded, slightly angulate at middle; distal margin deeply excised at middle, with the angles of the excision faintly produced. Saccus very broad, rounded.

Distribution: Central and South America.

This is one of the genera booked by Mr. TODD for a revision. Here, as in *Aucula* and *Darceta*, some species are not congeneric, e. g. *E. dolens* DRUCE and *E. trita* DRUCE. As said in the Introduction, the genus *Erocha* is being treated here in the usual way, so as not to interfere with Mr. TODD's forthcoming study.

Key to the species

1 Hind wings uniform dark brown with pale cilia 2
1′ Hind wings wholly or partly pale coloured 5
2 Fore wings dark reddish-brown with a white apical patch . . . . ***E. leucotelus***, p. 13
2′ Fore wings without white apical patch 3
3 Fore wings with the ground colour creamy white with basal and submarginal bronze-coloured markings . . . . ***E. picta*** ♂, p. 15

3′ Fore wings differently coloured 4
4 Fore wings with the costa broadly white . . . . *E. affinis*, p. 15
4′ Fore wings with the costa black sprinkled with white . . . . *E. irrorata*, p. 16
5 Hind wings blackish-brown with a small whitish discal patch . . . . *E. picta* ♀, p. 15
5′ Hind wings differently marked 6
6 Hind wings white or yellowish with broad dark margins 7
6′ Hind wings yellowish, without dark margins 10
7 Hind wings with the ground colour yellowish . . . . *E. elaeina*, p. 15
7′ Hind wings with the ground colour white 8
8 Hind wings with the white medial area partly sprinkled with black, and veins on it streaked with black . . . . *E. semiviridis*, p. 14
8′ Hind wings with the white medial area not sprinkled with black 9
9 Hind wings with veins streaked with black on the white medial area . . *E. mummia*, p. 12
9′ Hind wings with veins not streaked with black on the white medial area . . . . *E. leucodisca*, p. 13
10 Fore wings white, thickly sprinkled with chocolate-red . . . . *E. dolens*, p. 16
10′ Fore wings grey, very thickly sprinkled with black . . . . *E. trita*, p. 17

Erocha mummia (CRAMER)

(fig. 4)

1779 *Bombyx mummia* CRAMER, Uitlandsche Kapellen, 3:61; pl. 228c. – Terra typica: Suriname (Type not designated).
1864 *Dioptis discreta* WALKER, List Lepidopt. Het. Brit. Mus., 2:319. – Terra typica: Brazil, Amazonas, Ega. (Type, ♂, in British Museum).
1896 *Pycnodontis mummia* – JORDAN, Novit. Zool., 3:27.
1910 *Erocha mummia* – HAMPSON, Cat. Lepidopt. Phal. Brit. Mus., 9:409; fig. 193.

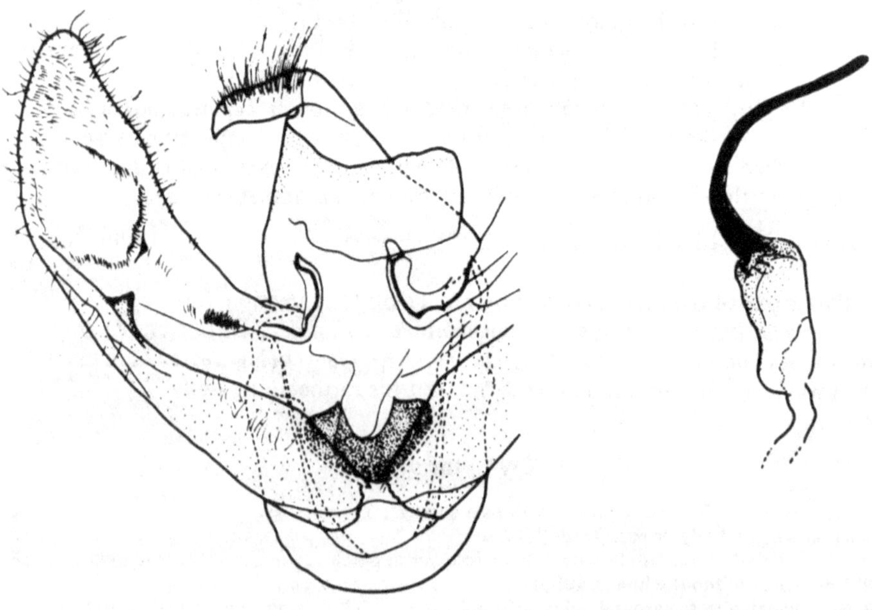

Fig. 4. *Erocha mummia* (CRAMER). BM Noct. 5144. Type

D i a g n o s i s : Head and thorax dark reddish-brown with some white hairs; basal antennal segment white; vertex with white spots above eyes; coxae white; abdomen blackish-brown; basal crest with some metallic scales; a lateral orange patch on urites II to IV. Fore wings dark reddish-brown, with olive tinge and some bluish-white sprinkling, especially on costal area; the media and bases of veins 3, 4, 6 and 7 slightly streaked with white; subbasal line represented by a rather obsolete dark striga from costa; antemedial line indistinct, oblique, slightly edged with whitish distally; cell-mark dark, very obscure; discocellular mark lunulate, dark, edged with white; postmedial line represented by a small white costal spot and an indistinct double dark line from the submedian fold to dorsum; a rather diffused dark brown line from costa beyond the postmedial line; an indistinct whitish subterminal line, edged distally by slight dark brown marks. Hind wings blackish-brown with a semi-hyaline white patch in, below and beyond the cell; the media and veins 2, 3 and 4 on it black; cilia white at the tips towards apex and tornus. Length of fore wing 18–20 mm.

D i s t r i b u t i o n : Venezuela, French Guiana, Suriname, Brazil.

Erocha leucodisca HAMPSON

1910 *Erocha leucodisca* HAMPSON, Cat. Lepidopt., Phal. Brit. Mus., 9:408; pl. 145, fig. 32. – T e r r a t y p i c a : Perù, Chanchamayo. (Type, ♂, in British Museum).

D i a g n o s i s : Antennae of ♂ pectinated, of ♀ serrated. Head and thorax chocolate-brown mixed with white hairs; pectus mostly white; abdomen brown mixed with some white, the fifth urite with an orange band; underside of the five basal urites yellow, the first urite with lateral tufts of long brownish ochreous hair. Fore wings chocolate-brown; area from costa to the submedian fold suffused with sap-green to the postmedial line except at base of the costal area; the media and veins from it streaked with white to the postmedial line; a large diffused white patch beyond cell from costa to vein 3, with irregular outer edge; subbasal line represented by a slight white striga from below costa; antemedial line represented by a slight double excurved line filled in with greenish from above vein 1 to dorsum; cell-mark represented by a slight brown spot with a white dot below it; discocellular mark represented by some greenish dots on its distal edge in interspaces; postmedial line represented by double dark spots on costa and a double sinuous line filled in with greenish from the submedian fold to dorsum; a rather diffused chocolate-brown line from costa just beyond the postmedial line; a whitish submarginal line; some white on termen in the submedian interspace. Hind wings white; costal and terminal areas broadly blackish-brown, the terminal band ending in a point on the anal margin before middle. Length of fore wing 25 mm.

D i s t r i b u t i o n : Perù.

Erocha leucotelus (WALKER)

1865 *Euthisanotia leucotelus* WALKER, List Lepidopt. Het. Brit. Mus., **32**:635. – T e r r a t y p i c a : Perú, Lima (Type, ♀, in British Museum).

1868/74 *Leiosoma serpentinum* FELDER, Reise Novara, p.6; pl. 99, fig. 23. – T e r r a t y p i c a : ignota (Type not designated).

1889 *Leiosoma juanita* SCHAUS, Entomologia Americana, 5:87. – Terra typica: Mexico, Paso de San Juan (Type in U.S. National Museum).
1896 *Leonides serpentinum* – DRUCE, Biol. Centr. Amer., Heterocera, 2: 328.
1910 *Erocha leucotelus* – HAMPSON, Cat. Lepidopt. Phal. Brit. Mus., 9:410.

Diagnosis: Head and thorax dark reddish-brown mixed with some white hair; fore tarsi banded with white; abdomen blackish-brown mixed with some whitish, with broad orange sublateral fasciae except at the extremity. Fore wings reddish-brown with an olive tinge; costal area more chocolate-brown to the postmedial line; costal, dorsal and terminal areas and the media sprinkled with bluish-white; discocellulars and base of veins 3 to 6 slightly streaked with white; an apical white patch to vein 6; subbasal line represented by an obscure dark striga from costa; antemedial line represented by indistinct double dark striae from costa and dorsum; postmedial line represented by a small white spot on costa and a double dark line from the submedian fold to dorsum; cell-mark as a very indistinct dark spot; discocellular mark defined by by a few bluish-white scales; a rather diffused dark chocolate-brown line from costa well beyond the wing-middle; subterminal line bluish-white, waved, from a white apical patch; a terminal series of small dark lunules edged by a waved bluish-white line except on the white area. Hind wings fuscous brown with slight reddish tinge; cilia pure white. Length of fore wing 21 mm.

Distribution: Perù, Colombia, Mexico.

Erocha semiviridis (DRUCE)

1903 *Euthisanotia semiviridis* DRUCE, Ann. Mag. Nat. Hist., (7) 11:200. – Terra typica: Perù, Cuzco (Type, ♂, in British Museum).
1910 *Erocha semiviridis* – HAMPSON, Cat. Lepidopt. Phal. Brit. Mus., 9:410; fig. 194.

Diagnosis: Head and thorax blackish-brown mixed with some white hairs; collar spotted with white; pectus mostly white; fore femora and tibiae fringed with brownish-white hair; abdomen blackish-brown mixed with whitish; underside with white segmental lines; lateral tufts of long orange-yellow hair from the base. Fore wings dull sap-green; the median vein from base, the discocellulars and bases of veins 3, 4, 5 and 7 streaked with white; a small white basal spot; subbasal line represented by a very oblique white striga from costa; antemedial and postmedial lines represented by indistinct double dark lines from the submedian fold to dorsum; the latter oblique; a large white patch from costa just beyond cell to vein 3 with irregular sinuous rather diffused edges; a diffused dark chocolate-brown line from costa well beyond the middle of wing, becoming from vein 3 a broad diffused fascia, bent inwards, below the submedian fold to base; a fine bluish-white subterminal line just beyond the brown line; white nervural streaks from it to termen towards apex; a fine bluish-white terminal line from vein 3 to tornus. Hind wings blackish-brown with a large white patch thickly sprinkled with black, in lower part of cell, below and well beyond the cell, veins on it streaked with black; cilia tipped with white. Length of fore wing 23 mm.

Distribution: Perù, Bolivia.

Erocha elaeina ZERNY

1916 *Erocha elaeina* ZERNY, Ann. Wiener Hofmuseum, **1916**:187; pl. 5, fig. 27. – Terra typica: Bolivia, Rio Songo (Type, ♂, in Vienna Museum).

Diagnosis: Body dark brown; a few white scales on head, palpi, thorax and legs. Fore wings dark violet grey, with sparse bluish sprinkling; cell-mark elliptical, dark olive-green, edged with dark brown; discocellular mark very large, nearly square, olive-green, edged with brown proximally, diffused distally; a broad olive-green stripe from base below cell, broadening to a large light brown patch rounded distally, edged with olive-green, reaching costa where it is sprinkled with yellowish-white above the discocellular mark, and with veins partly streaked with bluish-white in patch; latter edged distally with a broad, diffused dark chocolate-brown line, arched and slightly waved, broadening on costa to form a large spot, and forming a blunt angle between veins 4 and 5, and 5 and 6, ending in the median fold; in the dorsal area, the dark brown ground colour broadens towards the olive-green stripe, the broad portion limited proximally and distally by an ochreous vertical line from dorsum; cilia concolorous. Hind wings yellowish-white, with a sharply limited dark brown marginal band on costa, termen and anal margin; cilia dark brown. Length of fore wing 17–18 mm.

Distribution: Bolivia.

Erocha picta DRAUDT

1919 *Erocha picta* DRAUDT, in SEITZ, Großschmetterl. Erde, **11**:9; pl. 5B. – Terra typica: Bolivia, Coroico (Types, ♂♀, in Berlin Museum).

Diagnosis: Head and thorax blackish, sprinkled with bluish-white; abdomen black; anal tuft ochreous-yellow. Fore wings creamy-white; costal, dorsal and terminal areas black, thickly sprinkled with bluish-white scales, as also is the space between the median vein and vein 7, and the cell- and discocellular-marks; postmedial line represented by a wavy line encircling the discocellular mark; three bronzy clouds, viz. at base of costa, on dorsum and below the median vein, and seven oval marginal spots; cilia black, spotted basally with white. Hind wings black, with a white medial spot in ♀; a small yellow-orange spot at middle of costa. Length of fore wing 22 mm.

Distribution: Bolivia.

Erocha affinis DRAUDT

1919 *Erocha affinis* DRAUDT in Seitz, Großschmetterl. Erde, **11**:10; pl. 5b. – Terra typica: Bolivia, Rio Songo (Type, ♀, probably in Berlin Museum).

Diagnosis: Body black, mixed with white hair-scales on head and thorax; abdomen dorsally, and laterally on urites II to V, orange. Fore wings purplish-black, thickly sprinkled with violet-grey scales, which form before the termen an indistinct zigzag-line; middle of dorsum with a square spot of ground colour, variegated with black, and edged all around with olive-green; cell- and discocellular marks green, the latter with a medial white line; a large white postmedial spot, broadly margined with olive-green distally and dorsally; costal area broadly white; cilia white towards tornus. Hind wings blackish-brown, unspotted; cilia grey, whitish at tips. Length of fore wing 17 mm.

Distribution: Bolivia.

Erocha irrorata Jones

1915 *Erocha irrorata* JONES, Trans. entomol. Soc. London, **1914**:447. – Terra typica: S. E. Brazil, W. Sao Paulo, Araçatura. (Type, ♂, in British Museum).

Diagnosis: Head, collar, tegulae and thorax dark fuscous mixed with white; a white postocular mark; palpi dark brown, mixed with white and black; pectus white; legs fuscous, sprinkled with white; fore femora and tibiae with long white hair; abdomen dark fuscous with a lateral orange stripe. Fore wings dark fuscous sprinkled with white; central area dull green from near base below cell, along the submedian fold, in cell and on disk, the latter with a large white spot sprinkled with green from costa to vein 2; the green area limited by a narrow black diffused line; antemedial and postmedial lines green, visible only on dorsum, the former outcurved, confluent with the central green area on vein 1, the latter incurved, confluent with the green area on the submedian fold; costal streak black sprinkled with white; the median vein to vein 2, and a bar across cell at this point black sprinkled with white; an indistinct subterminal and a terminal line of white sprinkling; cilia fuscous. Hind wings dark fuscous; cilia white. Length of fore wing about 18 mm.

Distribution: Southeastern Brazil.

Erocha dolens (Druce)

1904 *Heterocampa dolens* DRUCE, Ann. Mag. Nat. Hist., (7)**13**:248. – Terra typica: S. E. Perù, Agualani (Type, ♂, in British Museum).
1910 *Erocha dolens* – HAMPSON, Cat. Lepidopt. Phal. Brit. Mus., **9**:407; fig. 142.

Diagnosis: Head and thorax mixed with white, black, and chocolate-red hairs and a few scales; metathorax with a band of white hair; tibiae fringed with white; abdomen black, yellow at the tip; metathoracic crest blue-black. Fore wings white, thickly sprinkled with chocolate-red; a white spot at the base of costa; antemedial line chocolate-red, interrupted in cell, edged proximally with a white band, diffused distally; cell- and discocellular marks chocolate-red edged with white except costally, the former round, the latter lunulate; postmedial line chocolate-red, edged distally by a white band; subterminal line represented by a bar from costa, then a lunulate line, the lunules touching termen; a fine black terminal line; cilia chequered fuscous and white. Hind yellowish-white; traces of a blackish postmedial line obliquely excurved from costa to vein 2; cilia white with a fex black scales at base. Length of fore wing 23 mm.

Distribution: Southeastern Perù, Bolivia, possibly Argentina.

Erocha trita DRUCE

1910 *Erocha trita* DRUCE, Ann. Mag. Nat. Hist., (8) **6**:178. – Terra typica: S. E. Perù, Agualani (Type, ♂, in British Museum).

Diagnosis: Head, abdomen and legs black; collar, tegulae and thorax grey; anal tuft brown. Fore wings grey, very thickly sprinkled with black scales; a round black spot in the cell; an oval-shaped spot at the end of the cell; a zigzag rather broad white line crosses the wing near the base from the costal to the dorsal margin; a second zigzag line crosses the wing beyond the cell, ending in a large white spot, close to the tornus. Hind wings pale cream-yellow, the marginal line black; the cilia white. Length of fore wing 22 mm (the original description mentions erroneously "expanse 2 in.").

Distribution: Southeastern Perù.

Genus **Aucula** WALKER

1882 *Aucula* WALKER, Trans. entomol. Soc. London, (3) **1**:253. – Species typica: *Aucula josioides* WALKER.
1896 *Arpia* SCHAUS, J. New York entomol. Soc. **4**: 148. – Species typica: *Arpia janeira* SCHAUS.

Diagnosis: Proboscis fully developed; antennae in male and female bipectinate, with rather short branches to the apex (in *A. hilzingeri* BERG and *A. magnifica* SCHAUS the branches are moderate in male, and the antennae serrate in female); palpi obliquely upturned, with the second joint reaching to about middle of frons, and fringed with long hair in front, and the last joint moderate and porrect; frons with a rounded prominence with a truncate conical process at centre, with raised edges and a corneous plate below it; eyes large, round.

Thorax hairy, without crests; tibiae fringed with long hair. Abdomen with rough hair, but without crests. Nervation: in fore wings, vein 2 from about seven-tenths of cell; 3, 4, 5 approximated, equidistant; discocellulars incurved; 6 from the upper quarter thereof; areole small, narrow; 7 and the short to moderate stalk of 8 + 9 from the extremity of the areole; 10 from its upper angle. In hind wings, vein 2 from about three-quarters of cell; 3, 4 and 6, 7 from the cell-angles; discocellulars incurved; 5 obsolescent, from middle of the discocellulars; 8 anastomosing with the cell at a point or for a very short distance.

Male genitalia: Uncus about one-half of sacculus in length, compressed, arched, pointed. Tubus analis broad, sclerified ventrally. Fold of vinculum broad, covered with sparse hair. Valva broad, especially basally; costa incurved for two-thirds, then angled outwards, and straight terminally; apex bearing a lobe covered with stiff hair; saccus very faintly and regularly outcurved; fold broad basally, narrowing and reaching to near the termen; a secundary fold from near the base, connected to the first fold, and bearing in its basal part a harpe directed caudad, shaped like the neck and head of some bird not unlike a diver. Aedeagus with process slightly longer than the costa, very robust, ending in a slender horn-shaped process; fultura inferior elongate, with the sides incurved and the proximal margin angled; fultura superior fused to fultura inferior, bearing lateral processes with a terminal plate connected with the valva, and with a few stiff hair on it. Saccus short, rounded. As has been pointed out in the Introduction, Mr. Todd is currently working on a revision of this genus. The latter certainly is but an assemblage of mostly not congeneric species, and, besides that, Mr. TODD has discovered in the British Museum at least twenty apparently new species. The present author has seen them in the collections of the British Museum, and they look heterogenous enough. Some species studied by the author also are not congeneric with the type-species, e. g. *Aucula buprasium* DRUCE and *A. sublata* (WALKER), the latter described as *Bepara sublata*. Understandably, the present author has abstained from any taxonomic conclusions pending Mr. TODD's investigations.

D i s t r i b u t i o n: Ecuador, French Guiana, Brazil, Bolivia, Argentina.

Key to the species

1 Antennae ciliate in both sexes 2
1′ Antennae bipectinate in both sexes or in male alone 3
2 Fore wings with a whitish costal fascia expanding triangularly before the middle, and with a hook beyond the middle . . . . *A. janeira*, p. 20
2′ Fore wings without a whitish costal fascia . . . . *A. sublata*, p. 21
3 Antennae bipectinate in both sexes 4
3′ Antennae ciliate in female 5
4 Hind wings with a broad orange patch, extending well below the submedian fold . . . . *A. josioides*, p. 18
4′ Hind wings with a narrow orange patch, not extending to below the submedian fold . . *A. buprasia*, p. 19
5 Abdomen with orange lateral patches at the base 6
5′ Abdomen without orange lateral patches basally . . . . *A. magnifica*, p. 20
6 In fore wings, the postdiscal line broad, complete; in hind wings, the black border extends to the base of the anal margin . . . . *A. hilzingeri*, p. 19
6′ In fore wings, the postdiscal line reduced to three white spots; in hind wings, the black border not extending to the anal margin . . . . *A. schausi*, p. 21

Aucula josioides WALKER

(fig. 5)

1862 *Aucula josioides* WALKER, Trans. Entomol. Soc. London, (3) 1:253. – T e r r a t y p i c a: Brazil (Type, ♀, in British Museum).

D i a g n o s i s: Head and thorax chocolate red mixed with white hairs; palpi and legs black mixed with white; fore coxae ochreous white; pectus in front orange; abdomen black. Fore wings chocolate red sprinkled with bluish white scales; a diffused yellowish patch beyond the cell extending to vein 2 and to the postmedial line, which is indistinct and oblique from costa to vein 6. Hind wings orange with a broad black terminal border, expanding at the costa to before middle, and on the anal margin to near base. Length of fore wing 26 mm.

D i s t r i b u t i o n: Brazil (Petropolis, Rio de Janeiro)

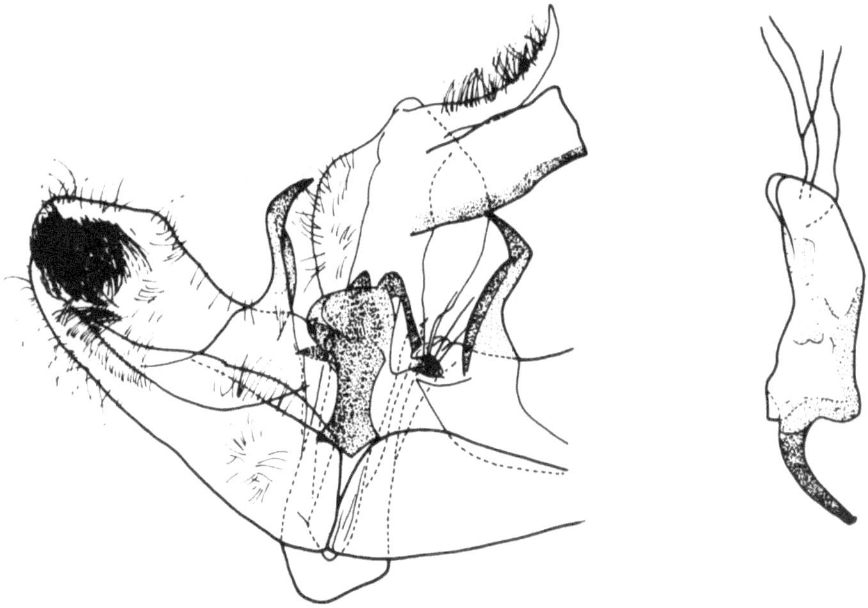

Fig. 5. *Aucula josioides* WALKER. BM Noct. 5164. Type

Aucula buprasia (DRUCE)

1897 *Leiosoma buprasium* DRUCE, Ann. Mag. Nat. Hist., (6) **20**:300. – Terra typica: Ecuador (Type, ♂, in British Museum).
1910 *Aucula buprasia* – HAMPSON, Cat. Lepidopt. Phal. Brit. Mus., **9**:420; fig. 199.

Diagnosis: Head and thorax deep chocolate red; antennae blackish brown; palpi and legs black sprinkled with white scales; coxae, femora and tibiae fringed with ochreous white hair; abdomen black with the anal tuft orange. Fore wings deep chocolate red sprinkled with bluish white scales on terminal half; a diffused yellowish patch beyond the cell, extending to vein 2 and to the postmedial line, which is indistinct, oblique from costa to vein 6, excurved to vein 3, then incurved. Hind wings orange; costal area to near the base, anal area to the submedian fold and a terminal band black, leaving the orange area rather narrow and arched above. Length of fore wing 19 mm.

Distribution: Ecuador, French Guiana, Bolivia.

Aucula hilzingeri (BERG)

1882 *Metagarista hilzingeri* BERG, Ann. Soc. Argent., **13**:176. – Terra typica: Argentina, Buenos Aires (Types, ♂♀, in Mus. Arg. Cienc. Nat.).
1910 *Aucula hilzingeri* – HAMPSON, Cat. Lepidopt. Phal. Brit. Mus., **9**:421; fig. 200.
1936 *Aucula hilzingeri* var. *albirubra* KÖHLER, Physis, **12**:73; pl. (middle row, left figure). – Terra typica: Argentina, Tucumàn (Type in coll. Breyer).

D i a g n o s i s : Head and thorax deep red mixed with white hair; pectus with some ochreous hair; abdomen blackish brown, with some orange at the base of sides and in the anal tuft. Fore wings deep red thickly sprinkled with white; a short curved white subbasal line; a red antemedial line angled outwards in the submedian fold where there is a white patch before it and a white streak from it to the postmedial line; mark on the discocellulars rather small, edged with red; postmedial line red, strongly bent outwards below the costa and angled outwards below vein 4, then retracted to below the cell-end and excurved to vein 1, with a white patch before and just beyond it from the costa to vein 3, and a lunule beyond it in the submedian interspace; an irregularly crenulate red line just before the termen, with white beyond it; a fine black terminal line; cilia chequered fuscous and white. Hind wings orange yellow with a black terminal band, oblique from the costa beyond the middle to vein 4, then narrow; cilia chequed black and white. Length of fore wing 21–26 mm.

D i s t r i b u t i o n : Argentina (Buenos Aires, Tucumàn)

Aucula magnifica (SCHAUS)

1904 *Euthisanotia magnifica* SCHAUS, Trans. Amer. entomol. Soc., **30**:150. – T e r r a t y p i c a : Brazil, Petropolis (Type, ♀, in U.S. National Museum).
1910 *Aucula magnifica* – HAMPSON, Cat. Lepidopt. Phal. Brit. Mus., 9:422; pl. 146, fig. 5.

D i a g n o s i s : Head and thorax blackish brown with some white hairs, especially on the head; collar with some deep red; fore tibiae with white streaks above; abdomen blackish brown with some white hairs on the underside. Fore wings deep chocolate red, thickly sprinkled with bluish white; subbasal line white, curved, from the costa to vein 1; antemedial line red, edged proximally by white except in the cell, waved from the cell to the dorsum; a short white streak beneath the cell; cell- and discocellular-marks black with a few white scales on them, the former round; a small v-shaped mark of white between the marks; the mark of the discocellulars edged with white distally and followed by a whitish patch; postmedial line red, edged distally with a white band except just below vein 2; a waved red submarginal line, edged by white on both sides, touching the termen in interspaces ii to vi; a fine black terminal line; cilia chequered white and black. Hind wings pale yellow; costal and terminal areas broadly, and anal margin chocolate blackish; a slight brown discocellular bar; slight terminal white streaklets at the veins; cilia as in fore wings. Length of fore wing 25 mm.

D i s t r i b u t i o n : Southeastern Brazil; Argentina (Tucumàn)

Aucula janeira (SCHAUS)

1896 *Arpia janeira* SCHAUS, J. New York entomol. Soc., **4**:48. – T e r r a t y p i c a : Brazil, Rio de Janeiro (Type in U.S. National Museum).
1897 *Copidryas marginalis* ROTHSCHILD, Novit. Zool., **4**:312; pl. 7, fig. 10. – T e r r a t y p i c a : Brazil, Espiritu Santo (Type in British Museum).
1910 *Aucula janeira* – HAMPSON, Cat. Lepidopt. Phal. Brit. Mus., **9**:422; fig. 201.

Diagnosis: Head grey; second palpal joint with a chocolate brown band; sides of frons chocolate; antennae black; thorax chocolate-red with a broad dorsal grey stripe, and grey stripes on the collar; pectus grey; legs blackish and grey; abdomen blackish brown; underside greyish towards the base. Fore wings chocolate red; costal area white, sprinkled with brown and chocolate-red, ending in a point before the apex, triangularly produced at middle of the cell to below the cell, connected with the dorsum by a fine oblique white line; another rounded projection beyond the cell, extending to interspace iii. Hind wings orange; costal and terminal areas broadly black; anal area black, narrowing to a point near the base. Length of fore wing 24 mm.

Distribution: Brazil (Rio de Janeiro)

Aucula sublata (WALKER)

1864 *Bepara sublata* WALKER, List. Lepidopt. Het. Brit. Mus., **31**:143. – Terra typica: Brazil, Amazonas, Ega (Type ♂ in British Museum).
1910 *Aucula sublata* – HAMPSON, Cat. Lepidopt. Phal. Brit. Mus., 9:423; pl. 146, fig. 16.

Diagnosis: Head and thorax deep chocolate-red mixed with whitish; abdomen blackish. Fore wings deep chocolate-red sprinkled with bluish-white; two indistinct antemedial lines, widely separated and oblique from costa to the submedian fold, then incurved and approximate; postmedial line indistinct, double, waved; subterminal line blue, dentate, from costa to vein 6, and a similar terminal line to vein 3. Hind wings orange yellow; costal and anal areas to near base and a terminal band black. Length of fore wing 15 mm.

Distribution: Brazil.

Aucula schausi JÖRGENSEN

1935 *Aucula schausi* JÖRGENSEN, Ann. Mus. Argent. Cienc. Nat., **38**:125–126; pl. 3, fig. 29. – Terra typica: Argentina, Catamarca, Los Cerros de Aconquija (Type ♂ in Museo Argent. Buenos Aires).

Diagnosis: ♂. Body black; legs, especially femora, with long white pilosity; first abdominal segments laterally orange; thorax and fore wings minium-red, sprinkled with white. In fore wings, antemedial line strongly diffused; cell-mark represented by a curved line; discocellular mark represented by two transverse arched lines, all minium-red, with white edging at near sides; postmedial line reduced to three white spots, viz., two costal; with broad minium-red proximal edges, and one, larger, on dorsum, proximally of tornus, and edged all around by minium-red; subterminal line represented by a series of minium-red spots; terminal line black, with white dots on the ends of veins and on the black cilia. Hind wings orange, with a black terminal band, broad costally and apically, but gradually becoming narrow towards the anal angle, and not present on the anal margin; the black spot hardly showing through from the underside. ♀ like ♂, but the cell- and discocellular marks more distinct; the two arched lines on the discocellular spot bear red spots, and fuse together to a single spot finely edged on both sides by a white line; there is a black discal point. Length of fore wing 23 mm.

Distribution: Argentina, Catamarca, Tucumàn.

Genus **Gerrodes** HAMPSON

1910 *Gerrodes* HAMPSON, Cat. Lepidopt. Phal. Brit. Mus., 9:418. – Species typica: *Diamuna longipes* DRUCE (by original designation).

Diagnosis: Proboscis fully developed; antennae ciliate and somewhat laminate; eyes large, round; palpi obliquely porrect to just beyond the frons; the second joint fringed with long hair in front; the third short. Frons with a long truncate conical projection with raised edges and a corneous plate below it. Thorax clothed with hair only; tegulae with thick tufts of upturned hair at the extremities; metathorax with a spreading crest; tibiae fringed with long hair. Abdomen with a dorsal basal crest only. Nervation: in fore wings, vein 2 from about three-quarters of cell; 3, 4, 5 approximated; distance 3, 4 about half of that between 4, 5; discocellulars incurved; 6 from the upper cell-angle; areole narrow, rhomboidal; 7 and the short stalk of 8 + 9 from the extremity of areole; 10 from near the extremity. In hind wings, vein 2 from about three-quarters of cell; 3, 4 and 6, 6 from the cell-angles; discocellulars well incurved; 5 obsolescent, from middle of discocellulars.

Male genitalia: Uncus nearly as long as half the costa, narrow, compressed, nearly straight, pointed at the extremity. Tubus analis short, broad, sclerified ventrally. Fold of vinculum rather narrow, bearing a bunch of moderate stiff hair. Valva rather broad, broadest at about one-quarter of sacculus; costa incurved in proximal half, then outcurved, finally nearly straight to the apex; the latter rounded; sacculus more or less parallel to costa; fold of sacculus slightly broader than half the breadth of valva, reaching to middle of the latter, and ending in a short, roughly triangular harpe, with its proximal edge toothed. Aedeagus slightly shorter than the sacculus, robust, straight, broadening and bilobate terminally; fultura inferior broad and short, with the proximal margin toothed at the middle and laterally; sides straight, faintly divergent; distal angles produced and covered with short spinulets. Saccus semi-elliptical.

Distribution: U.S.A., Mexico.

Key to the species

1 Fore wings leaden-grey, with a broad white band below the cell; outer line vertical, simple . . . . *G. minatea*, p. 24
1′ Fore wings bright reddish-brown, without a white dorsal band; outer line double below vein 3 . . . . *G. longipes*, p. 22

Gerrodes longipes (DRUCE)
(fig. 6)

1889 *Diamuna longipes* DRUCE, Biol. Centr. Amer., Heterocera, 1:334; pl. 30, fig. 18. – Terra typica: Mexico, Jalapa, Trujillo (Types, ♂♀, in British Museum).
1910 *Gerrodes longipes* – HAMPSON, Cat. Lepidopt. Phal. Brit. Mus., 9:419; fig. 198.

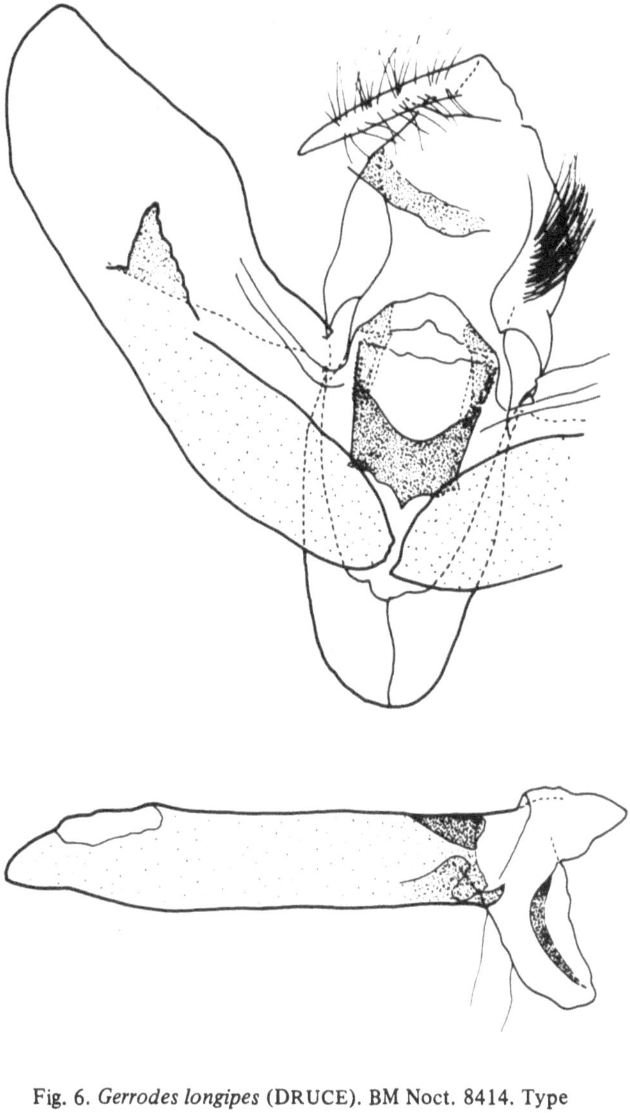

Fig. 6. *Gerrodes longipes* (DRUCE). BM Noct. 8414. Type

Diagnosis: Head and thorax reddish-brown, mixed with some grey; tuft at the extremity of the tegulae and the metathoracic crest brighter rufous; abdomen reddish-brown, suffused dorsally with blackish; the basal crest rufous. Fore wings bright reddish-brown; dorsal and terminal areas tinged with grey; a slight pale streak at the cell-end; pale streaks in interspaces beyond the upper cell-angle; a slight dark streak below basal half of the media; a minute subbasal streak above dorsum; proximal lines and cell- and discocellular marks absent; postmedial line dark brown, slightly edged with greyish distally, running from costa towards apex, excurved to vein 3, then double, bent inwards and running very obliquely to the submedian fold beyond the middle; a short white streak below it towards vein 2; subterminal line dark brown, edged with whitish proximally, excurved to vein 3, then bent inwards to vein 1 beyond the middle; a minutely waved dark terminal line; cilia pale rufous with a slight dark line through them. Hind wings orange with a broad blackish-brown terminal band expanding on the anal area; cilia greyish with a fine dark line through them. Length of fore wing 25–28 mm.

Distribution: Mexico.

Gerrodes minatea DYAR

1912 *Gerrodes minatea* DYAR, Proc. U.S. National Museum, **1912**:72. – Terra typica: Mexico, Zacualpan (Type, ♂, in U.S. National Museum).

Diagnosis: Head and thorax leaden grey with brown streaks; abdomen yellow with a dark dorsal stripe. Fore wings leaden grey streaked with brown; an oblique white shadow from before the apex into the cell, crossed by the brown streaks; outer line brown, parallel to termen in apical portion, then incurved below the cell and joining the other brown streaks; closely following it, a fine white line, also incurved and becoming a broad white band below the cell, not reaching the base; a slender whitish line beyond, in the grey terminal space; a leaden terminal line. Hind wings orange-yellow with a rather narrow blackish-brown terminal band; cilia whitish. Length of fore wing 24 mm.

Distribution: U.S.A. Mexico.

Genus **Xerociris** COCKERELL

1863 *Ciris* GROTE, Proc. entomol. Soc. Philadelphia, **2**:65 [nec C. L. KOCH 1850]. – Species typica: *Ciris wilsonii* GROTE (by original designation).
1904 *Xerociris* COCKERELL, Entomol. News, Philadelphia, **15**:76 (nomen novum pro *Ciris* GROTE nec C. L. KOCH).

Diagnosis: Proboscis fully developed; palpi moderately long, obliquely porrect with the second joint clothed with appressed scales, and the third moderate; frons with a large conical pointed prominence; antennae of male bipectinate, with rather long branches to apex; eyes large, rounded; collar and thorax chiefly scaled; tegulae clothed with long hair; prothorax with a divided haircrest; metathorax with a spreading crest; abdomen with a large crest on the basal urite. Nervation: in fore wings, vein 2 from about two-thirds of cell; 3, 4, 5 approximated, equidistant; discocellulars well incurved; areole narrow; 6 from areole; 7 and the short stalk of 8 + 9 from the extremity of areole; 10 from the upper angle of areole. In hing wings, vein 2 from slightly before five-eighths of cell; 3, 4 from the lower cell-angle (or 3 sometimes slightly before angle); discocellulars incurved; 5 obsolescent, from middle of discocellulars; 6, 7 from the upper cell-angle.

Male genitalia: Uncus moderate, about one-third of sacculus in length, compressed, very faintly arched, ending in a sharp point; tubus analis short and broad, nearly wholly sclerified. Fold of vinculum rather narrow, partly covered with long hair.

Valva rather broad, upcurved, with costa and sacculus nearly parallel; apex seemingly incurved; fold of sacculus broad basally, narrowing distally, reaching to middle of valva and ending in a large lobe bearing the small, thorn-shaped harpe. Aedeagus relatively very large, slightly longer than valva, moderately robust in the proximal two-thirds, spoon-shaped in the proximal fourth, arched and becoming slenderer distally, bearing at the beginning of the bend a dorsal and a ventral patch of minute spine-like cornuti; fultura inferior bilobate proximally, fused distally to the fultura superior, and forming lateral squarish lobes edged laterally and distally with rows of minute spinulets. Saccus rather angulate semi-circular.

Distribution: U.S.A., Texas.

Xerociris wilsonii (GROTE)

(fig. 7)

1863 *Ciris wilsonii* GROTE, Proc. entomol. Soc. Philadelphia, **2**:65; pl. 3, fig. 1. – Terra typica: Western Texas, Shovel Mounts (Type in U.S. National Museum).
1904 *Xerociris wilsonii* – COCKERELL, Entomol. News, Philadelphia, **15**:76.
1910 *Xerociris vilsoni* – HAMPSON, Cat. Lepidopt. Phal. Brit. Mus., **9**:400–401; fig. 189.

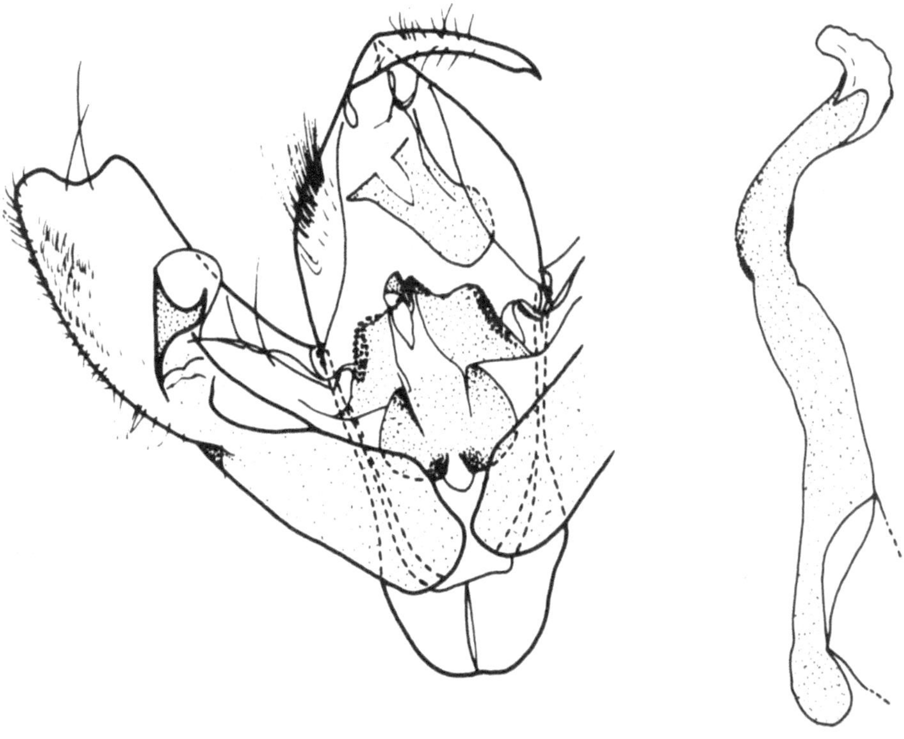

Fig. 7. *Xerociris wilsonii* (GROTE). BM Noct. 8410

D i a g n o s i s : Head and thorax blackish brown, mixed with bluish-black; antennae brown; tegulae pure white; legs white; tarsi marked with brown on joints; abdomen white; dorsal crests blackish-brown. Fore wings silvery grey with grey subbasal spots below costa and cell; a blackish fascia mixed with olive-green on the costal area from one-sixth to three-quarters, connected with the elliptical cell-spot and the large fabiform discocellular mark, which both are olive-green; an olive-green antemedial bar with a silvery purplish-grey patch on it from dorsum to the submedian fold, continued along dorsum as a fascia to the terminal band; latter deep purplish-grey with some white dentate marks, and with a sinuous olive-green proximal edge; cilia purplish-grey with two pale lines. Hind wings pure white, with a chocolate-red patch on termen from tornus to vein 2, and a point above vein 2; cilia white, but purplish-grey at the red patch. Length of fore wing 18–20 mm.

E a r l y s t a g e s : Larva: Head reddish brown with a black spot on the face of each lobe; body robust, enlarged at joint 12, purplish black, each ring with a transverse red patch subventrally, and a complete band on joint 12; several narrow transverse white lines on each segment; on thorax and joint 12 several of the setae are very long, flattened and spatulate; shields and legs red, tipped with black. Feeds on Vinum.

D i s t r i b u t i o n : U.S.A (Texas).

Genus **Orthia** HERRICH-SCHÄFFER

1850 *Orthia* HERRICH-SCHÄFFER, Außereurop. Schmetterl., :13. – S p e c i e s t y p i c a : *Orthia augias* HERRICH-SCHÄFFER (by monotypy).
1877 *Othria* WESTWOOD, Trans. Linn. Soc., Zool., (2) 1:201. – S p e c i e s t y p i c a : *Othria amazonica* WESTWOOD = *Orthia augias* HERRICH-SCHÄFFER).

D i a g n o s i s : Proboscis fully developed; antennae strongly dilated towards the extremity; palpi upturned, fringed with hair in front; the third joint short, with rough hairscales; frons with a truncate conical prominence, with a raised rim at the extremity; tibiae fringed with rough hair above; first urite clothed dorsally with rough hair. Nervation: fore wings with vein 2 from five-sixths of cell; 3 and 4 from the lower cell-angle; 5 well above the angle; upper discocellular well incurved; 6 from just below the upper cell-angle; areole minute, sometimes absent; 7, 10, 8 + 9 from the extremity of areole; stalk of 10, 8 + 9 short, that of 8 + 9 long. In hind wings, vein 2 from three-quarters of cell; 3, 4 and 6, 7 from the cell-angles; discocellulars slightly angled; 5 obsolescent, from middle of discocellulars.

Male genitalia: Uncus about seven-tenths of costa in length, compressed, strongly bent basally, nearly straight for most of its length, ending in a sharp point. Tubus analis broad, with dorsal sclerifications at the base. Fold of vinculum narrow, without hair. Valva rather broad; costa nearly straight; apex well rounded; sacculus outcurved, more distinctly so just beyond the middle; fold of sacculus rather narrow, especially distally, ending in a short, sickle-shaped harpe. Aedeagus relatively very large, about as long as valva, very robust, spoon-shaped in proximal half, club-shaped distally, with an oblique subterminal patch of small cornuti; fultura inferior with the proximal margin angled outwards, the sides converging, the terminal margin evenly incurved; fultura superior with the lateral margins sclerified and forming a lobe-shaped fold at the base of valva. Saccus rather elongate semi-elliptical.

Distribution: South America.

Orthia augias HERRICH-SCHÄFFER
(fig. 8)

1850 *Orthia augias* HERRICH-SCHÄFFER, Außereurop. Schmetterl., fig. 18. – Terra typica: South America (Type not designated).
1874 *Agarista lethe* FELDER, Reise Novara, : pl. 107, fig. 7. – Terra typica: Not designated.
1874 *Eusemia lethe* – BOISDUVAL, Rev. Mag. Zool., (3) **2**:102.
1875 *Orthia nexa* BOISDUVAL, Spec. Gen. Lepidopt. Het., **1**:543. – Terra typica: Not designated.
1877 *Othria nexa* – WESTWOOD, Trans. Linn. Soc., Zool., (2) **1**:201.
1877 *Othria amazonica* WESTWOOD, Trans. Linn. Soc., Zool., (2) **1**:202. – Terra typica: Amazonas, Pebas (Type, ♀, in Oxford Museum).
1892 *Othria lethe* – KIRBY, Syn. Cat. Lepidopt. Heterocera, :32.
1901 *Othria augias* – HAMPSON, Cat. Lepidopt. Phal. Brit. Mus., **3**:655–656; fig. 290.

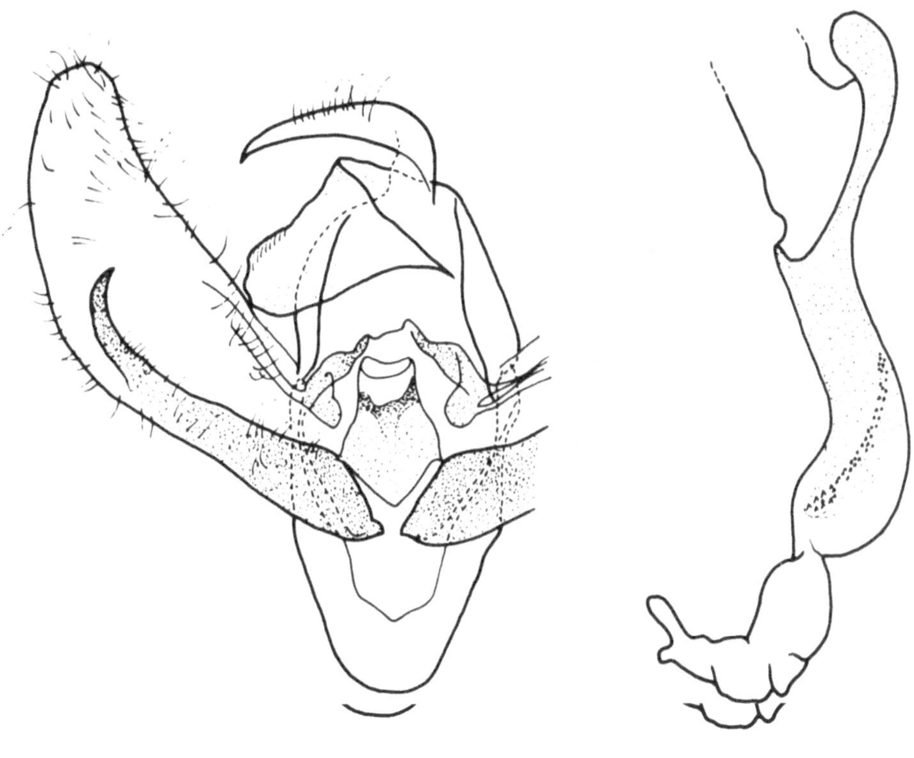

Fig. 8. *Orthia augias* HERRICH–SCHÄFFER. BM Noct. 8404. Type

D i a g n o s i s : Body and wings black; fore coxae and femora orange laterally; wings slightly shot with blue. Fore wings with an oblique orange or yellow band from just below costa slightly beyond middle to just above tornus, with somewhat irregular edges, and varying in shape. Hind wings unspotted. Length of fore wing 23–26 mm.

D i s t r i b u t i o n : Colombia, Bolivia, the Amazon region, Perù, Argentina.

Genus **Seirocastnia** GROTE

1866 *Seirocastnia* GROTE, Proc. entomol. Soc. Philadelphia, 6 :175. – S p e c i e s t y p i c a : *Ephialtias tribuna* HÜBNER (by original designation).

D i a g n o s i s : Proboscis fully developed; antennae slightly dilated towards the extremity; palpi upturned, fringed in front with long hair; last joint minute, clothed with rough hair; frons bearing a tuft of hair and a small truncate prominence with slightly raised rim at the extremity; tibiae fringed with rough hair. Nervation: fore wings with vein 2 from slightly beyond three-quarters of cell; 3, 4, 5 well separated, distance between 3 and 4 slightly longer than distance between 4 and 5; discocellulars incurved; 6 from the upper cell-angle; areole very small, triangular; 7, 8 + 9 very short stalked from the extremity of areole, very closely to the origin of 10; stalk of 8 + 9 long. Hind wings with vein 2 from five-sevenths of cell; 3, 4 and 6, 7 from the cell-angles; discocellulars faintly incurved; 5 obsolescent, from middle of discocellulars.

Male genitalia: Uncus relatively small, slightly less than half of costa in length, arched basally, then nearly straight, pointed. Tubus analis rather broad, sclerified dorsally. Fold of vinculum rather narrow, clothed distally with long stiff hair. Valva moderately broad, upturned, with costa and sacculus roughly parallel, although the latter is slightly more strongly outcurved in middle portion; fold of sacculus rather broad basally, narrowing to reach slightly before one-third of sacculus, and bearing terminally a short arched harpe; a submarginal apical row of semi-stiff setae present. Aedeagus slightly shorter than costa, robust, spoon-shaped for slightly more than half of its length proximally; vesica bearing a sclerified plate with two small patches of rudimentary cornuti; fultura inferior with the proximal margin sharply angled and produced at middle; sides slightly converging, then parallel; distal margin very faintly incurved; fultura superior broad, with lateral margins narrowly sclerified. Saccus short semi-elliptical.

D i s t r i b u t i o n : West Indies, Central and South America.

Key to the species

1 Abdomen with some orange or yellow, at least laterally 2
1′ Abdomen wholly black 8
2 Hind wings with costa yellow at base 3
2′ Hind wings with costa wholly black 4
3 Hind wings with anal margin orange . . . . *S. amalthea*, p. 29
3′ Hind wings with anal margin black nearly to base . . . . *S. panamensis*, p. 29
4 Abdomen with narrow dorsal black stripe; hind wings with costa and anal margin narrowly black 5
4′ Abdomen with broad black dorsal stripe; hind wings with costa and anal margin broadly black 6

5 Fore wings with an oblique orange yellow postmedial band narrowing to a point below vein 2 near termen . . . . *S. volupia*, p. 30
5′ Fore wings with an oblique yellow postmedial band expanded at middle and rounded above tornus . . . . *S. extensa*, p. 32
6 Fore wings with an orange streak in cell . . . . *S. tribuna*, p. 30
6′ Fore wings without orange streak in cell 7
7 Hind wings with the black costal fascia not extending to cell . . . . *S. nocturn, a*, p. 31
7′ Hind wings with the black costal fascia extending to cell . . . . *S. meridiana*, p. 31
8 Fore wings with an oblique orange band . . . . *S. columbina*, p. 32
8′ Fore wings with a white postmedial band . . . . *S. albifascia*, p. 32

Seirocastnia amalthea (DALMANN)

1823 *Aegocera amalthea* DALMANN, Anal Entomol., :49. – Terra typica: South America (Type not designated).
1856 *Josia separata* WALKER, List Lepidopt. Het. Brit. Mus., 7:1645. – Terra typica: Colombia, Bogota (Types, ♂♀, in British Museum).
1874 *Agarista lindigii* FELDER, Reise Novara, : pl. 107, fig. 6. – Terra typica: South America (Type not designated).
1874 *Eusemia lindigii* – BOISDUVAL, Rev. Mag. Zool., (3) 2:103.
1877 *Othria amalthea* – WESTWOOD, Trans. Linn. Soc., Zool., (2) 1:202
1877 *Othria lindigii* – WESTWOOD, Trans. Linn. Soc., Zool., (2) 1:202.
1892 *Phasis amalthea* – KIRBY, Syn. Cat. Lepidopt. Heterocera, : 32.
1892 *Phasis separata* – KIRBY, Syn. Cat. Lepidopt. Heterocera, : 32.
1901 *Seirocastnia amalthea* – HAMPSON, Cat Lepidopt. Phal. Brit. Mus., 3:657.

Diagnosis: Head, thorax and fore wings black; collar orange in front; pectus and legs with orange pilosity; abdomen orange with dorsal black stripe, lateral lines and black extremity. Fore wings with a medial orange mark, very variable, sometimes strongly reduced to a small triangular patch (typical form), to an oblique elliptical spot from cell to above vein 1 (*separata*), or extended to a large triangular patch reaching on vein 1 to near base and beyond middle (*lindigii*); a postmedial oblique band from vein 6 to interval ii, pointed at the extremities, also very variable. Hind wings orange, with a broad black terminal band. Length of fore wing 22 mm.

Distribution: Ecuador, Colombia.

Seirocastnia panamensis HAMPSON

1884 *Phasis separata* DRUCE, Biol. Centr. Amer., Heterocera, 1:36 [nec *Josia separata* WALKER 1856]. – Terra typica: Panamà, Chiriqui (Type, ♂, in British Museum).
1901 *Seirocastnia panamensis* HAMPSON, Cat. Lepidopt. Phal. Brit. Mus., 3:657; pl. 53, fig. 16 (nomen novum nec species nova).

Diagnosis: Head, thorax and fore wings black; collar orange in front; pectus with orange pilosity; femora fringed with orange hair; abdomen orange, with dorsal black stripe and slight lateral streaks, and the extremity black. Fore wings with a triangular orange antemedial spot and an oblique orange postmedial band from below costa to below vein 2 near the termen. Hind wings orange, with a broad black terminal band, running inwards on costa and anal margin. In male the antemedial patch on the fore wings is absent. Length of fore wing 22--27 mm.

Distribution: Panamà.

Seirocastnia volupia (DRUCE)

1897 *Othria volupia* DRUCE, Ann. Mag. Nat. Hist., (6) **20**:299. – Terra typica: Santa Lucia, Chonana (Type, ♂, in British Museum).
1901 *Seirocastnia volupia* – HAMPSON, Cat. Lepidopt. Phal. Brit. Mus., **3**:657; pl. 53.

Diagnosis: Head, thorax and fore wings black; base of palpi, pectus, streaks on femora and on tibiae yellow; abdomen yellow with narrow dorsal and lateral black stripes and the extremity black. Fore wings with an oblique orange-yellow psotmedial band from below costa to below vein 2 near termen where it narrows to a point. Hind wings orange; costa narrowly black to base; anal area suffused with black, and narrowly margined with black; a rather broad terminal black band, expanding widely on tornal area. Length of fore wing 26 mm.

Distribution: West Indies.

Seirocastnia tribuna (HÜBNER)

(fig. 9)

1819 *Ephialtias tribuna* HÜBNER, Verzeichn. bek. Schmetterl., **11**:176. – Terra typica: Cuba (Type not designated).
1856 *Josia continua* WALKER, List Lepidopt. Het. Brit. Mus., **7**:1646. – Terra typica: Haiti (Type, ♀, in British Museum).
1866 *Seirocastnia tribuna* – GROTE, Proc. entomol. Soc. Philadelphia, **6**:175.
1884 *Phasis tribuna* – DRUCE, Biol Centr. Amer., Heterocera, **1**:36.
1887 *Eusemia schausii* HY. EDWARDS, Entomologia Americana, **3**:90. – Terra typica: Vera Cruz (Type in Amer. Mus. Nat. Hist.).
1892 *Phasis schausii* – KIRBY, Syn. Cat. Lepidopt. Heterocera, :33.

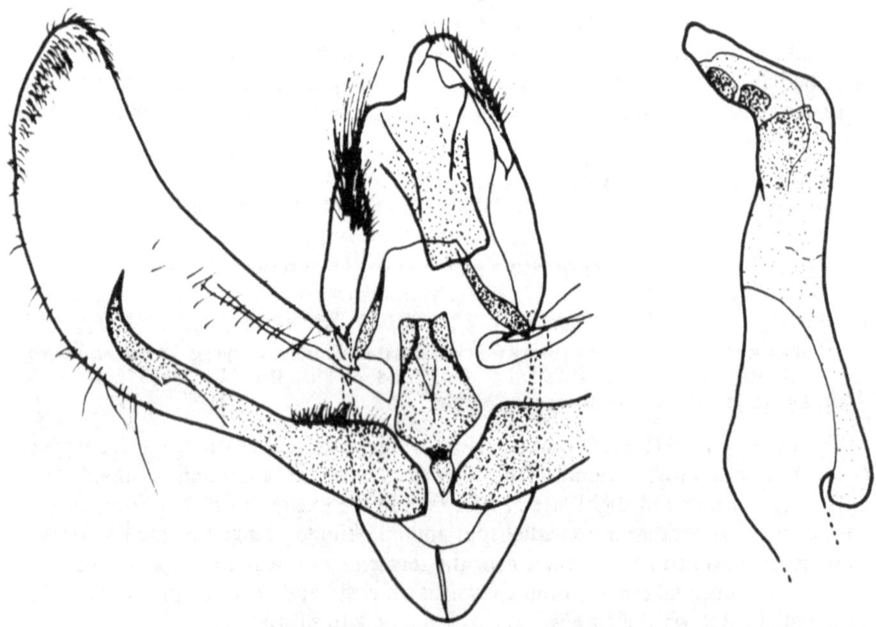

Fig. 9. *Seirocastnia tribuna* (HÜBNER). BM Noct. 8406

D i a g n o s i s : Head, thorax and wings black; base of palpi orange yellow; collar orange in front; pectus orange; mid and hind femora and tibiae fringed with orange hair; abdomen orange, with dorsal and ventral black stripes, and the anal tuft black. Fore wings with an orange wedge-shaped cell-streak, varying in width and length; an oblique band just below costa above the cell-angle to vein 2 near termen, also variable in width and shape. Hind wings with an orange fascia from base to before termen on vein 4, narrowing towards both extremities. Length of fore wing 24 mm.

D i s t r i b u t i o n : Mexico, Honduras, Guatemala, Nicaragua, Cuba, Haiti.

Seirocastnia nocturna BURMEISTER

1879 *Seirocastnia nocturna* BURMEISTER, Descr. Rep. Argent., Atlas, :64. – T e r r a t y p i c a : Argentina (Type not designated).
1896 *Phasis praefecta* DRUCE, Ann. Mag. Nat. Hist., (6) **18**:28. – T e r r a t y p i c a : Bolivia (Type, ♂, in British Museum.)
1897 *Orthia elaphebolia* DRUCE, Ann. Mag. Nat. Hist., (6) **20**:299. – T e r r a t y p i c a : Ecuador (Type, ♂, in British Museum).
1901 *Seirocastnia praefecta* – HAMPSON, Cat. Lepidopt. Phal. Brit. Mus., **3**:658; pl. 53, fig. 4.
1912 *Seirocastnia praefecta* aberr. *nervalis* – STRAND, Lepidopt. Cat., **5**:39.

D i a g n o s i s : Head, thorax and wings blackish brown; collar with a more or less developed orange band in front; coxae and femora with some orange hair; abdomen orange, with a black dorsal line and lateral lines; extremity and ventral segmental lines black. Fore wings sometimes with an orange medial spot on the median vein (*nervalis*); a narrow postmedial oblique orange band from below costa to vein 2 near termen, obsolescent (typical form) or distinct (*elaphebolia*). Hind wings with an orange patch from base to a point on vein 4 near termen, extending at middle from just below costa to vein 1. Length of fore wing 22–26 mm.

D i s t r i b u t i o n : Bolivia, Ecuador, Perù, Argentina (Tucumán).

Seirocastnia meridiana (SCHAUS)

1896 *Phasis meridiana* SCHAUS, J. New York entomol. Sco., **4**:148. – T e r r a t y p i c a : Brazil, Sao Paulo (Type, ♂, in U.S. National Museum).
1901 *Seirocastnia meridiana* – HAMPSON, Cat. Lepidopt; Phal. Brit. Mus., **3**:659; pl. 54, fig. 17.

D i a g n o s i s : Head, thorax and wings black; abdomen yellow with black dorsal and lateral stripes and the extremity black. Fore wings with a narrow rather irregular oblique orange band beyond cell between veins 2 and 7. Hind wings with a pear-shaped orange patch in, below and beyond cell, rounded at the extremity. Length of fore wing 25 mm.

D i s t r i b u t i o n : Brazil.

Seirocastnia columbina (WESTWOOD)

1877 *Othria columbina* WESTWOOD, Trans. Linn. Soc., Zool., (2) **1**:202. – Terra typica: Colombia (Type in Oxford Museum).
1892 *Phasis columbina* – KIRBY, Syn. Cat. Lepidopt. Heterocera,:32.

Diagnosis: Body and wings black; palpi with an orange spot near base; collar with an orange band in front. Fore wings with an oblique orange band from just below costa at the upper cell-angle to just above tornus, varying somewhat in width and shape. Hind wings unspotted. Length of fore wing 26 mm.

Distribution: Colombia.

Seirocastnia extensa JORDAN

1908 *Seirocastnia extensa* JORDAN, Novit. Zool., **15**:255. – Terra typica: Colombia, Rio Dagua (Type in British Museum).

Diagnosis: Head, thorax and fore wings black; base of palpi and pectus yellow; abdomen yellow with a slight dorsal black line and the anal tuft black. Fore wings with a broad oblique postmedial yellow band from below costa to above tornus, expanding at middle and rounded at the dorsal extremity. Hind wings yellow with a black terminal band extending narrowly on costa and anal margin to near base. Length of fore wing 30 mm.

Distribution: Colombia

Seirocastnia albifascia JOICEY & TALBOT

1922 *Seirocastnia albifascia* JOICEY & TALBOT, Bull. Hill Mus, **1**:301–302. – Terra typica: French Guiana, St. Laurent du Maroni (Type, ♀, in British Museum).

Diagnosis: Body fuscous black; first joint of palpi yellowish white laterally; legs fuscous black; fore legs except tarsi fringed with yellow hair; underside of abdomen paler. Fore wings blackish brown; a broad white band from costa to the submedian fold, with its distal edge curved and more sharply defined than the proximal edge which crosses the bases of intervals iii and vi and borders the discocellulars; the band narrows below vein 2 and is rounded at its lower extremity close to the tornus. Hind wings blackish brown with a deep blue sheen, unmarked. Length of fore wing 28 mm.

Distribution: French Guiana.

Genus Epithisanotia gen. nov.

Species typica: *Eudryas sancti-johannis* STEPHENS (designation hoc loco).

Diagnosis: Proboscis fully developed; antennae simple in both sexes (ciliate in ♂ of *Euthisanotia*); palpi upturned, the second segment longer than the third and covered with long scales; the last segment clothed with short scales. Tibiae without hair-tufts. Abdomen with dorsal crests on urites I and II. Fore wings with termen faintly angulated on vein 3; hind wings rather short, broad and well rounded, but with costa straight. Neuration: In fore wings, vein 2 arising slightly before three-quarters of cell; 3, 4 and 5 well separated; distance between 3 and 4 about twice that between 4 and 5; discocellulars short, well incurved; areole short and narrow; vein 6 arising from the upper angle of the discocellulars; 7 and 8 + 9 from the extremity of areole, stalk of 8 + 9 about one-third of 8; 10 from upper side of areole, near its extremity. In hind wings, vein 2 given off at four-fifths of cell; 3 and 4 much approximated; discocellulars incurved; 5 weak, from middle of discocellulars; 6 and 7 from a point; 8 approximated to cell at the extreme base (in *Euthisanotia*, vein 2 more proximal in both wings, veins 3 and 4 in hind wings from a point, and vein 8 fused to cell for a while near the base).

Male genitalia: Uncus short, compressed, arched, with a minute terminal hook. Tubus analis very broad, sclerified both dorsally and ventrally. Vinculum rather broad, well fringed with long hair. Valva rather narrow, with the apex upturned; sacculus slightly more convex, with the fold short and rather broad ending in a short, tolerably broad harpe which bears a small terminal clawlike process. Aedeagus slightly shorter then the valva, rather robust, spoon-shaped proximally, and slightly swollen in the terminal third, with a ventral row of minute spinules; fultura inferior constisting in a small proximal plate with concave distal margin, and lateral lobes distally narrowing and covered with minute spinules; fultura superior membranous, with the sides broadened, sclerified and covered with similar spinules. Saccus roughly triangular.

Distribution: Mexico.

Epithisanotia sanctijohannis (STEPHENS)

(fig. 10)

1850 *Eudryas sancti-johannis* STEPHENS, Brit. Animals in Collections Brit. Mus., **5**:312. – Terra typica: Horsley Down, England (Type, ♂, in British Museum).
1856 *Eudryas sanctae-johannis* WALKER, List Lepidopt. Het. Brit. Mus., 9:144 (ex errore pro sancti-johannis).
1892 *Euthisanotia sanctae-johannis* – KIRBY, Syn. Cat. Lepidopt. Heterocera,:39.

Diagnosis: Head and thorax reddish-brown mixed with white; palpi and frons blackish, the former white at the base; tegulae with white patches; pectus and base of legs orange yellow; rest of legs reddish-brown mixed with some white; tarsi ringed with white; abdomen orange with dorsal brown spots on basal half; underside whitish. Fore wings white; basal half suffused with reddish-brown, leaving a white fascia from base of costa and thence below cell; some yellow suffusion above vein 1 except at the extremity of the dark area; cell-spot red, edged with white, elliptical; discocellular spot olive, angled inwards on the median vein; post-medial line very distal, brownish-red with an olive band on the proximal side; area

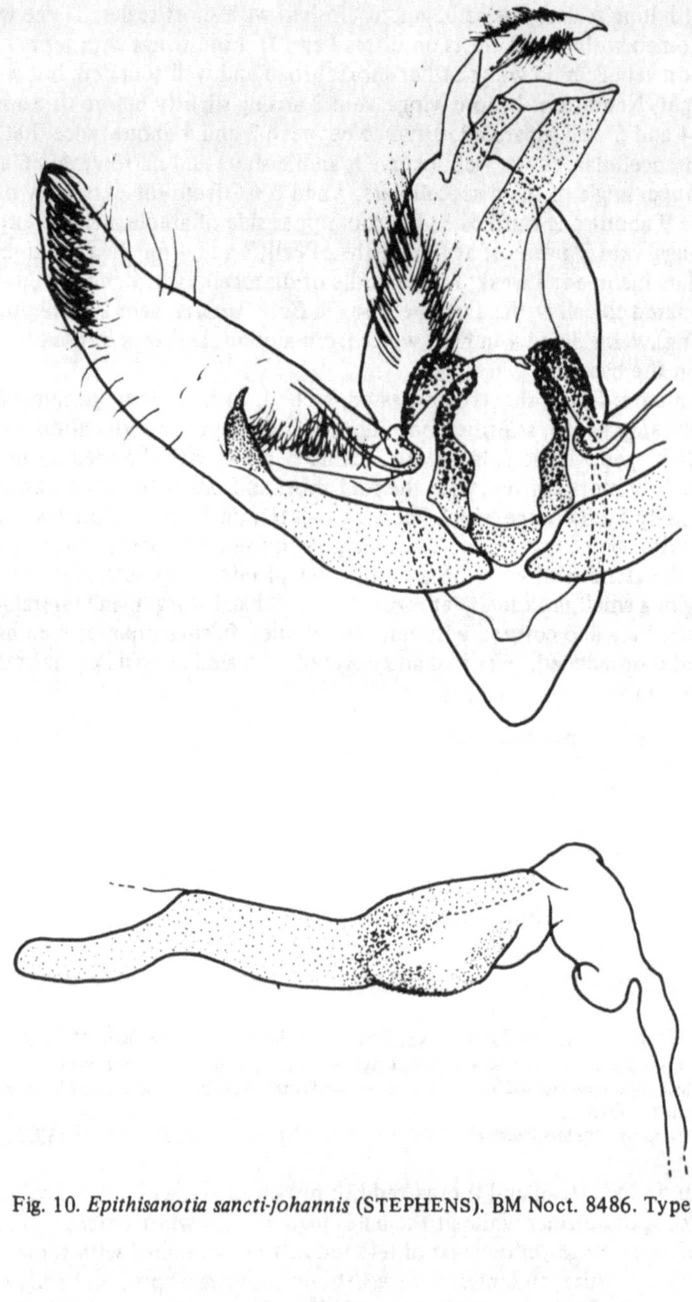

Fig. 10. *Epithisanotia sancti-johannis* (STEPHENS). BM Noct. 8486. Type

beyond indian red with traces of a waved whitish subterminal line. Hind wings orange, unspotted and unmarked. Length of fore wing 23 mm.

D i s t r i b u t i o n : England (Horsley Down, accidental); Mexico.

Genus **Darceta** HERRICH-SCHÄFFER

1855 *Darceta* HERRICH-SCHÄFFER, Außereurop. Schmetterl.,:13. – Species typica: *Darceta hesperina* HERRICH-SCHÄFFER = *Phalaena prosperpina* CRAMER (by monotypy).
1857 *Diamuna* WALKER, List Lepidopt. Het. Brit. Mus., **12**:960. – Species typica: *Phalaena severa* CRAMER (by monotypy).
1857 *Clitis* WALKER, List Lepidopt. Het. Brit. Mus., **12**:961. – Species typica: *Phalaena proserpina* CRAMER, (by monotypy).

Diagnosis: Proboscis fully developed; antennae slightly dilated towards the extremity; palpi oblique, roughly scaled, with the last joint short; frons with a truncate conical projection, with a raised rim at the extremity; tibiae fringed with long hair; abdomen with a dorsal crest on the first urite. Apex of the fore wings slightly falcate; in ♂, a fringe of long black hair on the underside of the fore wings above the inner margin, and a fringe of long yellow hair on the upperside of the hind wings below the basal half of costa. Venation: in fore wings, vein 2 arising from about three-quarters of the cell; 3, 4 and 5 well separated, equidistant; upper part of the discocellulars oblique, incurved; 6 from the upper angle of the cell; areole small and short; vein 7 stalked with the short stalk of 8 + 9, from the extremity of the areole; 10 from the upper angle thereof. In hind wings, vein 2 arising from about the middle of the cell; 3, 4 and 6, 7 from the cell-angles; discocellulars incurved; 5 obsolescent, from the middle of the discocellulars.

Male genitalia: Uncus rather short, narrow, ending in a minute hook. Tubus analis short and broad. Vinculum with a well marked fold, fringed with stiff hair. Valva elongate, narrow basally, broad in the middle portion, and slightly narrowing towards the apex; sacculus fringed with dense hair in its middle portion; harpe short. Aedeagus about one-half of the valva in length, slender, faintly arched; vesica bearing a patch of minute cornuti. Saccus semi-elliptical.

Distribution: Central and South America.

As said in the Introduction, Mr. TODD and Dr. FRANCLEMONT are planning to revise this genus; the present author has accordingly abstained from taxonomic changes in his treatment, although they seem inevitable as a result of the study of the male genitalia. However, it should be pointed out here, that *Darceta severa* (CRAMER), which is one of the species showing "aberrant" male genitalia, is the type-species of the genus *Diamuna* WALKER. The latter name accordingly is available for it and possibly for other species should it be advisable to remove the latter from the genus *Darceta*.

Key to the species

1 Fore wings with the costa lobed and curled over above at middle, with a depressed elliptical patch of ribbed membrane below it; apical area very truncate, and veins very strongly downcurved 2
1′ Fore wings normal; a fringe of long black hair on underside, above the anal margin . . . 3
2 Fore wings with a narrow yellowish line from base to middle; inner line oblique from cell to dorsum . . . . *D. falcata*, p. 38
2′ Fore wings without yellowish basal line; inner line oblique from middle costa to near base . . . . *D. haenschi*, p. 39
3 Hind wings with the anal area black 4
3′ Hind wings with the anal area yellow 5

4 Abdomen black; fore wings sprinkled with bluish-white and variegated with olive-brown; no triangular velvety red-brown spot below cell . . . . *D. proserpina*, p. 36
4′ Abdomen fuscous brown; fore wings sprinkled with lilacine on costa and termen; a triangular velvety red-brown spot below cell . . . . *D. grandimacula*, p. 37
5 Abdomen yellow . . . . *D. ophideres*, p. 37
5′ Abdomen dark brownish-black, at least above 6
6 Abdomen yellow on underside . . . . *D. primulina*, p. 37
6′ Abdomen blackish on underside . . . . *D. severa*, p. 39

Darceta proserpina (CRAMER)

(fig. 11)

1782 *Phalaena (Noctua) proserpina* CRAMER, Uitheemsche Kapellen, 4 239; pl. 399 I. – Terra typica: Suriname (Type not designated).
1856 *Darceta hesperina* HERRICH-SCHÄFFER, Außereurop. Schmetterl., 1 :13. – Terra typica: Bolivia, Rio Songo.
1856 *Darceta hesperica* – HERRICH-SCHÄFFER, Außereurop. Schmetterl.,: fig. 524–526.
1857 *Clitis proserpina* – WALKER, List Lepidopt. Het. Brit. Mus., 12:961.
1901 *Diamuna proserpina* – HAMPSON, Cat. Lepidopt. Phal. Brit. Mus., 3:653; fig. 287.
1901 *Diamuna hesperina* – HAMPSON, Cat. Lepidopt. Phal. Brit. Mus., 3:653.

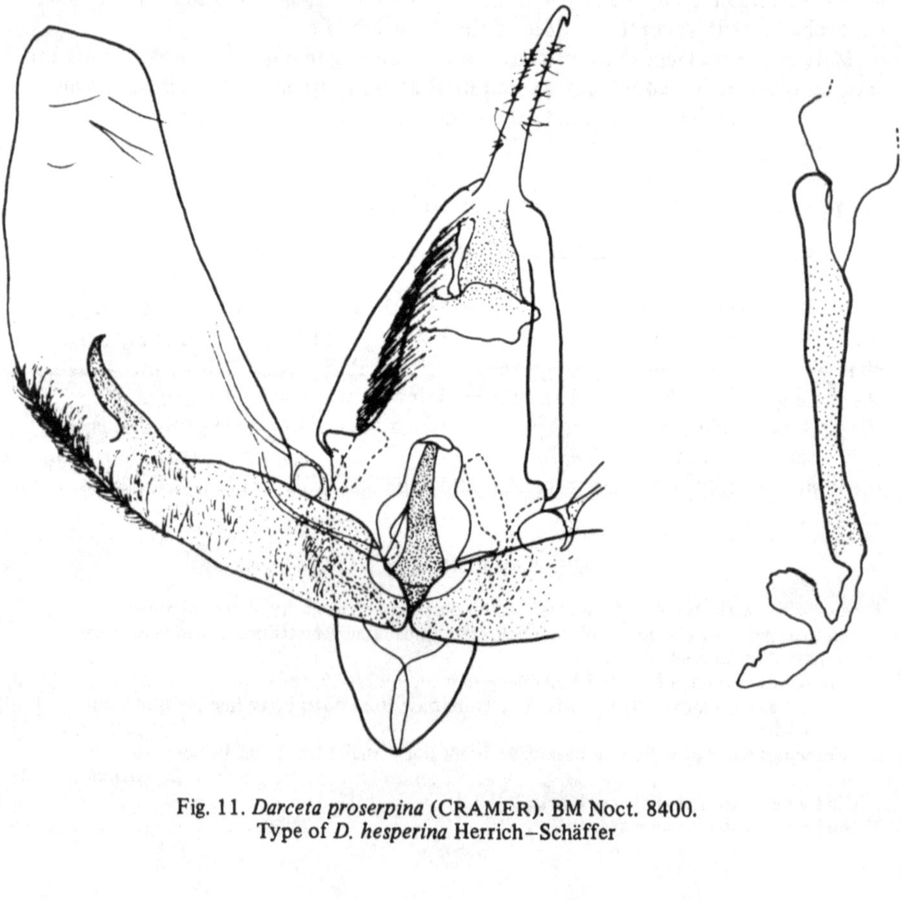

Fig. 11. *Darceta proserpina* (CRAMER). BM Noct. 8400.
Type of *D. hesperina* Herrich–Schäffer

D i a g n o s i s : Head and thorax deep reddish-brown; pectus and legs with some fulvous hair; abdomen black. Fore wings deep reddish-brown sprinkled with bluish-white; an oblique medial olive-brown band from cell to dorsum, rounded above; a postmedial olive-brown line on a white-sprinkled ground, edged on each side by chocolate-brown, strongly excurved from below costa to vein 3, then very oblique and incurved. Hind wings black, with a large elliptical orange patch beyond cell from below vein 8 to vein 2. Length of fore wing 22–25 mm.

D i s t r i b u t i o n : Panamà, Bolivia, Brazil, Colombia, Suriname, possibly Argentina.

Darceta primulina (DRUCE)

1889 *Clitis primulina* DRUCE, Biol. Centr. Amer., Heterocera, 1:335; pl. 30, fig. 23. – T e r r a t y p i c a : Mexico Presidio (Type, ♂, in British Museum).
1901 *Diamuna primulina* – HAMPSON, Cat. Lepidopt. Phal. Brit. Mus., 3:652.
1919 *Darceta primulina* – DRAUDT, in SEITZ, Großschmetterl. Erde, 11:6; pl. 1c.

D i a g n o s i s : Head and thorax yellowish-brown; abdomen dark brown; pectus, legs and underside of abdomen orange; fore tibiae fringed with blackish hair. Fore wings yellowish-brown; traces of an antemedial line from cell to dorsum; an indistinct fine curved subterminal line; terminal area greyish-brown. Hind wings orange-yellow with a broad black terminal band expanding rather irregularly below vein 2. Length of fore wing 19 mm.

D i s t r i b u t i o n : Mexico.

Darceta ophideres DRAUDT

1919 *Darceta ophideres* DRAUDT, in SEITZ, Großschmetterl. Erde, 11:7; pl. 5f. – T e r r a t y p i c a : Costa Rica, Orosi (Type, ♀, in Berlin Museum).

D i a g n o s i s : Head and thorax chocolate-brown; abdomen yellow. Fore wings chocolate-brown; dorsal half darker; above the submedian fold a fine streak of silvery-blue scales; silvery-blue scaling with violet reflections forming a rather indistinct streak enclosing the discocellulars and reaching the postmedial band; cell- and discocellular marks dark, the latter divided by a fine silvery line; postmedial line somewhat lighter, waved and edged on each side by darker. Hind wings yellow. Length of fore wing 23 mm.

D i s t r i b u t i o n : Costa Rica.

Darceta grandimacula (SCHAUS)

1921 *Diamuna grandimacula* SCHAUS, Insecutor Institiae Menstruus, 9:52. – T e r r a t y p i c a : S. E. Brazil, Joinville (Type, ♂, in U.S. National Museum).

D i a g n o s i s : Head and palpi dark reddish-brown; collar and thorax chestnut-brown; tegulae tipped with fuscous; fore tibiae and base of tarsi with long dark brown and black tufts; mid- and hind femora clothed with orange-brown hairs;

tibiae brown and fuscous; tarsi black; abdomen fuscous brown; basal tuft chestnut-brown; anal tuft fuscous-grey; underside similar, with long sublateral tufts. Fore wings dark purple tinged with brown; a fine black streak through cell to postmedial line; a triangular antemedial dark reddish-brown velvety spot below cell, its base resting on the media, followed at cell by some pale reddish-brown shading and a dark vertical streak; a fuscous streak at base of dorsum followed by lighter shading; a dull brown strak on the costal margin; a faint indication of an oval costal spot, followed by some lilacine sprinkling; postmedial line placed distally, brownish-black, outcurved from costa to vein 4, then slightly incurved to dorsum, edged distally with cinnamon-brown, and closely followed by a subterminal dark reddish line interrupted from vein 4 to below vein 6; a reddish-brown streak from subterminal line to apex below costa; termen tinged with lilacine; cilia brown. Hind wings black; a large yellow patch before termen, narrowly incurved along the costal margin to base, its inner edge straight from upper cell-angle to vein 2 near termen, its outer edge rounded. Length of fore wing 22 mm.

Distribution: Southeastern Brazil.

Darceta falcata (DRUCE)

(fig. 12)

1883 *Hecatesia falcata* DRUCE, Biol. Centr. Amer., Heterocera, 1:35; pl. 5, fig. 23–24. – Terra typica: Panamà, Bugaba (Types, ♂♀, in British Museum).
1901 *Diamuna falcata* – HAMPSON, Cat. Lepidopt. Phal. Brit. Mus., 3:651–652; fig. 286.
1919 *Darceta falcata* – DRAUDT, in SEITZ, Großschmetterl. Erde, 11:6; pl. 1b, 5f.

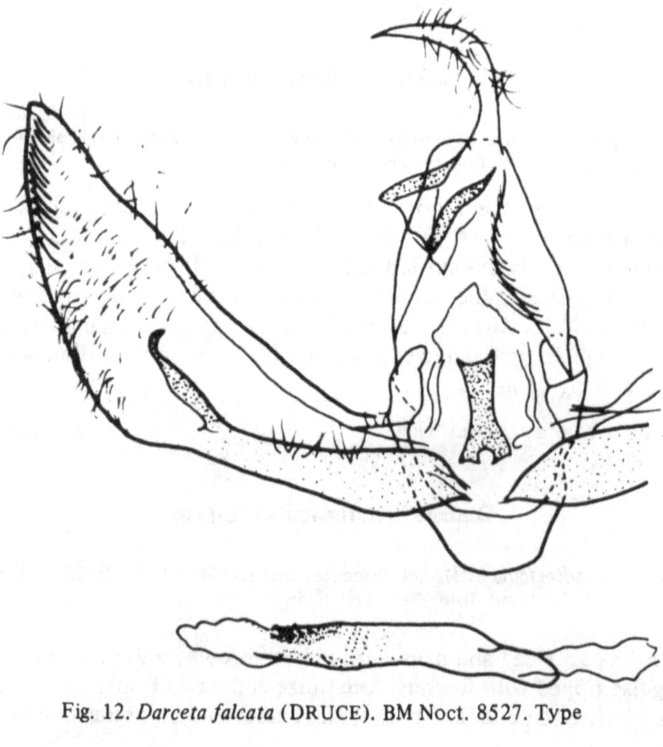

Fig. 12. *Darceta falcata* (DRUCE). BM Noct. 8527. Type

D i a g n o s i s : Body and fore wings deep red-brown; head, thorax and fore legs with some white hairs; pectus, mid- and hind femora and tibiae orange; abdomen orange; dorsal basal crest rufous. Fore wings in ♂ sprinkled with a few bluish-white scales forming slight streaks on some of the veins; in ♀ sprinkling much stronger; a yellow-brown antemedial mark below costa; a double oblique line from cell to dorsum with two elongate spots beyond it; a double postmedial line strongly angled beyond the lower cell-angle, then oblique in ♂, more erect in ♀, and with a diffused silvery band beyond it; veins of terminal area slightly streaked with yellow-brown. Hind wings orange, with a very broad blackish-brown terminal band, its inner edge angled at middle; cilia white at tip below apex in ♀. Length of fore wing 15 mm ♂, 21 mm ♀.

D i s t r i b u t i o n : Panamà.

Darceta haenschi DOHRN

1906 *Darceta haenschi* DOHRN, Stett. entomol. Zeitschr., **67**:275–276, text-fig. – T e r r a t y p i c a : W. Ecuador, Palmar (Type originally in Stettin Museum, possibly removed to Warsaw post 1945).

D i a g n o s i s : Body and wings deep red-brown; abdomen orange with the basal crest brownish. Fore wings as in *D. falcata*, but without the narrow yellowish line from base to the middle of wing; antemedial line nearly straight, oblique from middle of costa to near base; postmedial line running from four-fifths of costa to five-sixths of dorsum, strongly angled beyond the lower cell-angle. Hind wings dark brown with a rounded orange-yellow spot at cell-end, and a narrow streak on the anal margin. Length of fore wing 19 mm.

D i s t r i b u t i o n : West Ecuador.

Darceta severa (CRAMER)

(fig. 13)

1782 *Phalaena severa* CRAMER, Uitheemsche Kapellen, **4**:235; pl. 398 L. – T e r r a t y p i c a : Suriname (Type not designated).
1857 *Diamuna severa* – WALKER, List Lepidopt. Het. Brit. Mus. **12**:960.
1919 *Darceta severa* – DRAUDT, in Seitz, Großschmetterl. Erde, **11**:6; pl. 1 c.

D i a g n o s i s : Body and fore wings dark reddish-brown; pectus, mid- and hind femora and tibiae orange; abdomen black. Fore wings slightly sprinkled with silvery-blue; traces of an antemedial erect line below cell; postmedial line rather indistinct, double, edged by bluish distally, excurved from costa to vein 3, then incurved. Hind wings deep orange with a black terminal band with curved inner edge. Length of fore wing 25 mm.

D i s t r i b u t i o n : Mexico to Suriname, Bolivia, Amazonas, possibly Argentina.

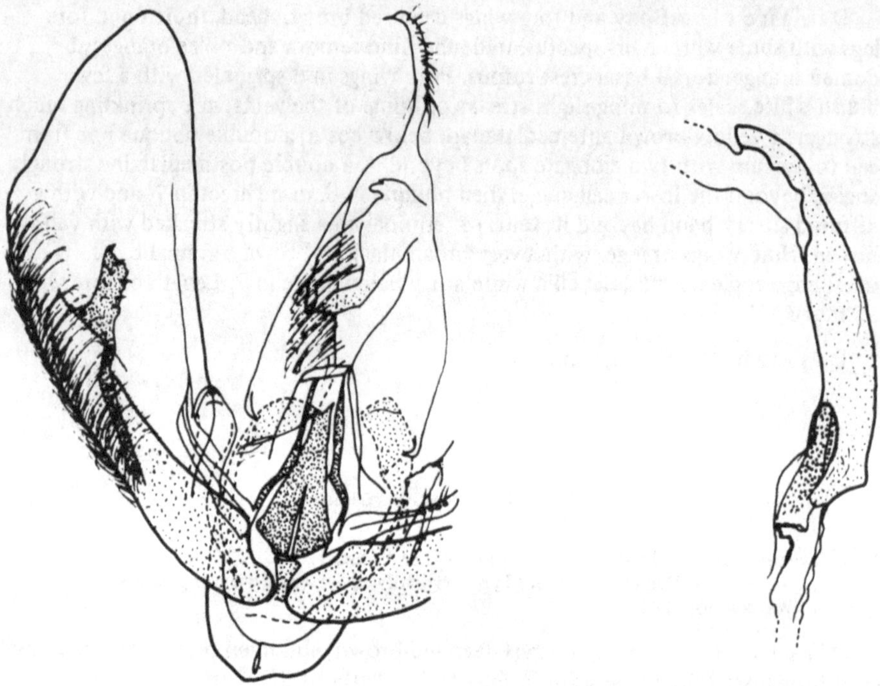

Fig. 13. *Darceta severa* (CRAMER). BM Noct. 8401

Genus **Copidryas** GROTE

1876 *Copidryas* Grote, Canad. Entomologist, **8**:99. – Species typica: *Euschirropterus gloveri* GROTE & ROBINSON (by original designation).

Diagnosis: Proboscis full developed; palpi obliquely upturned, fringed with hair; the last joint moderate, roughly scaled; frons with a large, flattened, hollowed-out corneous process with three horns, reaching slightly beyond the palpi; antennae slightly dilated beyond middle; tibiae unspined, but with long hair; abdomen with dorsal crests on the basal segments. Nervation: In fore wings, vein 2 from two-thirds of cell; 3, 4, 5 separate, equidistant; discocellulars incurved; areole narrow; 6 from the upper cell-angle; 7 and the stalk of 8 + 9 from the tip of areole; 10 from about three-quarters of areole on the upperside. In hind wings, vein 2 from five-sixths of cell; 3, 4 and 6, 7 from the cell-angles; discocellulars incurved; 5 obsolescent, from middle of discocellulars.

Male genitalia: Uncus slender, slightly longer than half of costa, compressed distally, forming an acute angle at two-sevenths of its length, then nearly straight, with a minute terminal hook. Tubus analis broad, sclerified ventrally. Vincula rather narrow, with their folds bearing a broad bunch of stiff hair. Valva very elongate and narrow, banana-shaped; terminal quarter of sacculus with a marginal row of short stiff hairs; fold of sacculus as broad as the valva at the base, then narrowing, and ending in a short curved harpe placed slightly beyond the first third of the valva. Aedeagus slightly shorter than the costa, spoon-shaped proximally, rather slender distally, nearly straight; a subterminal narrow field of cornuti; fultura inferior with a square proximal projection, narrowing distally, with the terminal margin excised at middle; median portion membraneous; fultura superior broadly lozenge-shaped, with lateral sclerifications. Saccus semi-elliptical, slightly pointed.

Distribution: U.S.A., Mexico.

Key to the species

1 Hind wings with a yellow terminal spot above tornus . . . . *C. gloveri*, p. 41
1' Hind wings without yellow terminal spot . . . . *C. cosyra*, p. 42

Copidryas gloveri (GROTE & ROBINSON)

(fig. 14)

1868 *Eusschirropterus gloveri* GROTE & ROBINSON, Trans. amer. entomol. Soc., 2:185. – Terra typica: Arizona (Type in American Museum of Natural History).
1876 *Copidryas gloveri* – GROTE, Canad. Entomologist, 8:99.

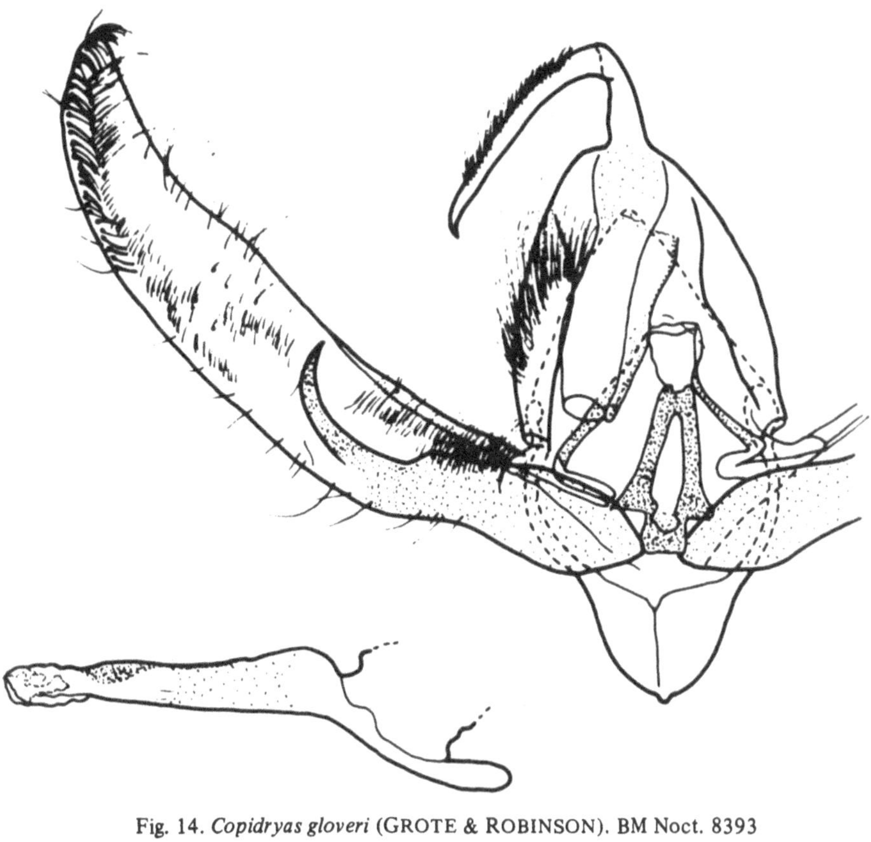

Fig. 14. *Copidryas gloveri* (GROTE & ROBINSON). BM Noct. 8393

D i a g n o s i s : Head and thorax mixed dark brown and grey hairs; extremity of the 2nd palpal joint and sides of frons white; tarsi ringed with white; abdomen orange; dorsal crests and anal tuft above black. Fore wings dark chocolate strongly sprinkled with grey; a white fascia from base of the costa in and below cell, its upper edge indented by the cell- and discocellular marks; these with dark edges and leaden centres; fascia then bent up to the costa before the apex; an oblique silvery-leaden antemedial striga from the fascia to vein 1, and an oblique interrupted post-medial line; a terminal series of black streaklets on a whitish ground. Hind wings orange-yellow; a dark brown terminal band with a yellow terminal spot above the tornus; cilia white except towards the tornus. Length of fore wing 18–20 mm.

Early stages: Larva: Young: pale or yellowish green with dark shading across middle of body. Adult; pale grey or dull white with black dashes on the side of each segment, and with shadings of salmon-pink. Feeds on Portulaca oleracea. Pupa: in a tubular burrow with gummed lining and closed with particles of earth.

D i s t r i b u t i o n : U.S.A.; Mexico.

Copidryas cosyra (DRUCE)

1896 *Euthisanotia cosyra* DRUCE, Biol. Centr. Amer., Heterocera, **2**:328; pl. 52, fig. 4. – T e r r a t y p i c a : Trujillo, Mexico (Type, ♀, in U.S. National Museum).
1910 *Copidryas cosyra* – HAMPSON, Cat. Lepidopt. Het. Brit. Mus., Suppl. **2**:583.
1938 *Misa cosyra* – MC DUNNOUGH, Check List Lepidopt. Canada USA, **1**:54.

D i a g n o s i s : Head and thorax mixed with dark red brown and grey; abdomen orange with the underside white. Fore wings dark brown thickly sprinkled with grey and with pale olive scales above and below the white fascia which begins just below costa near the base, then widens and extends from just above the media to the submedian fold to the cell-end, then bent upwards and narrowing to a point on the costa before the apex, dilated beyond the lower cell-angle, its dorsal edge strongly excised between the submedian fold and vein 3 and slightly between veins 3 and 4; discocellular mark indistinct, edged with dark; a silvery blue line on the distal side of the fascia from the costa to vein 5; a waved dark terminal line, slightly edged with white towards the tornus. Hind wings orange; a dark brown terminal band, angled on veins 1 and 2, and not reaching the tornus; a waved orange subterminal line, obsolescent towards the apex; cilia white, blackish towards the tornus. Length of fore wing 20 mm.

D i s t r i b u t i o n : Mexico; United States (Arizona).

Genus **Euscirrhopterus** GROTE

1866 *Euscirrhopterus* GROTE, Proc. entomol. Soc. Philadelphia, **6**:176. – S p e c i e s t y p i c a : *Euscirrhopterus poeyi* GROTE (by monotypy).
1866 *Heterandra* HERRICH-SCHÄFFER, Corresp.-Blatt zool. mineral. Verein Regensburg, **20**:134. – S p e c i e s t y p i c a : *Heterandra disparilis* HERRICH-SCHÄFFER = *Euscirrhopterus poeyi* GROTE
1892 *Euschirropterus* KIRBY, Syn. Cat. Lepidopt. Heterocera,:40.
1896 *Laquea* JORDAN, Novit. zool., **3**:41. – S p e c i e s t y p i c a : *Euthisanotia argentata* DRUCE = *Euscirrhopterus poeyi* GROTE, (by original designation).

D i a g n o s i s : Proboscis fully developed; antennae strongly dilated towards the extremity; palpi upturned, fringed with long hair; last joint moderate; frons with conical truncate prominence; tibiae fringed with hair; tarsi strongly spined and hairy; abdomen with dorsal crests on the basal urites. Venation: in fore wings, vein 2 given off at five sixths of cell; 3, 4 and 5 well separated, equidistant; upper portion of discocellulars incurved; areole small, much narrower in ♀; 6 from the areole, near its base; in ♂, 7 from underside of the areole, or 7, 10 and the stalk of 8, 9 from the extremity of the areole; in ♀, 8 and the very short stalk of 9 and 10 from the extremity of the areole; in ♂, costa strongly lobed and dilated before apex. In hind wings, vein 2 given off at five sixths of cell; 3, 4 and 6, 7 from a point at the cell-angles; discocellulars incurved; 5 obsolescent, from middle of the discocellulars.

Male genitalia: Uncus slightly longer than half of costa, slender, compressed, well arched in the proximal third, then faintly so, ending in a small hook. Tubus analis narrow at the base, then broadening, with dorsal and ventral sclerifications in the basal portion. Vincula broad; fold narrow, fringed with long, stiff hair. Valva very elongate and narrow, banana-shaped; apex pointed; terminal three tenths of sacculus bearing a row of stiff marginal hairs; fold of sacculus as broad as the valva at the base, but short, not reaching one-third of the fold of the sacculus, and ending in a rather short, sickle-shaped harpe. Aedeagus about one-half of the costa in length, spoon-shaped proximally, rather robust, but less so terminally, bearing terminally a marginal patch of minute thorn-shaped cornuti; fultura inferior narrow, membranous in its middle portion; fultura superior broad, squarish, with lateral sclerifications. Saccus square with the angles rounded.

D i s t r i b u t i o n : West Indies, Central and South America.

Key to the species

1 Frons with a pointed conical prominence; in fore wings, vein 10 coincident with 11 in ♂, strongly anastomosing with it in ♀ . . . . *E. walkeri*, p. 45
1′ Frons with a truncate conical prominence; in fore wings, vein 10 not coincident with 11 . . 2
2 Fore wings with an oblique discocellular spot; hind wings with the terminal band rather narrow . . . . *E. poeyi*, p. 43
2′ Fore wings with a lunulate discocellular spot; hind wings with the terminal band broad . 3
3 Fore wings with a large semicircular greyish-white patch along most of the costa; cell without oval brown spot . . . . *E. disciferus*, p. 46
3′ Fore wings with a whitish streak expanding to costa near the base; cell with an oval brown spot . . . . *E. klagesi*, p. 45

Euscirrhopterus poeyi GROTE

(fig. 15)

1866 *Euscirrhopterus poeyi* GROTE, Proc. entomol. Soc. Philadelphia, **6**:178. – T e r r a t y p i c a : Cuba (Type in U.S. National Museum).
1866 *Heterandra disparilis* HERRICH-SCHÄFFER, Corresp.-Blatt. zool. mineral. Verein Regensburg, **20**:134. – T e r r a t y p i c a : Cuba (Type in Berlin Museum).
1892 *Euschirropterus poeyi* – KIRBY, Syn. Cat. Lepidopt. Heterocera,:40.
1894 *Euthisanotia argentata* DRUCE, Ann. Mag. Nat. Hist., (6) **13**:352. – T e r r a t y p i c a : Mexico, Jalapa Orizaba (Type, ♀, in British Museum).
1896 *Laquea argentata* – JORDAN, Novit. zool., **3**41.

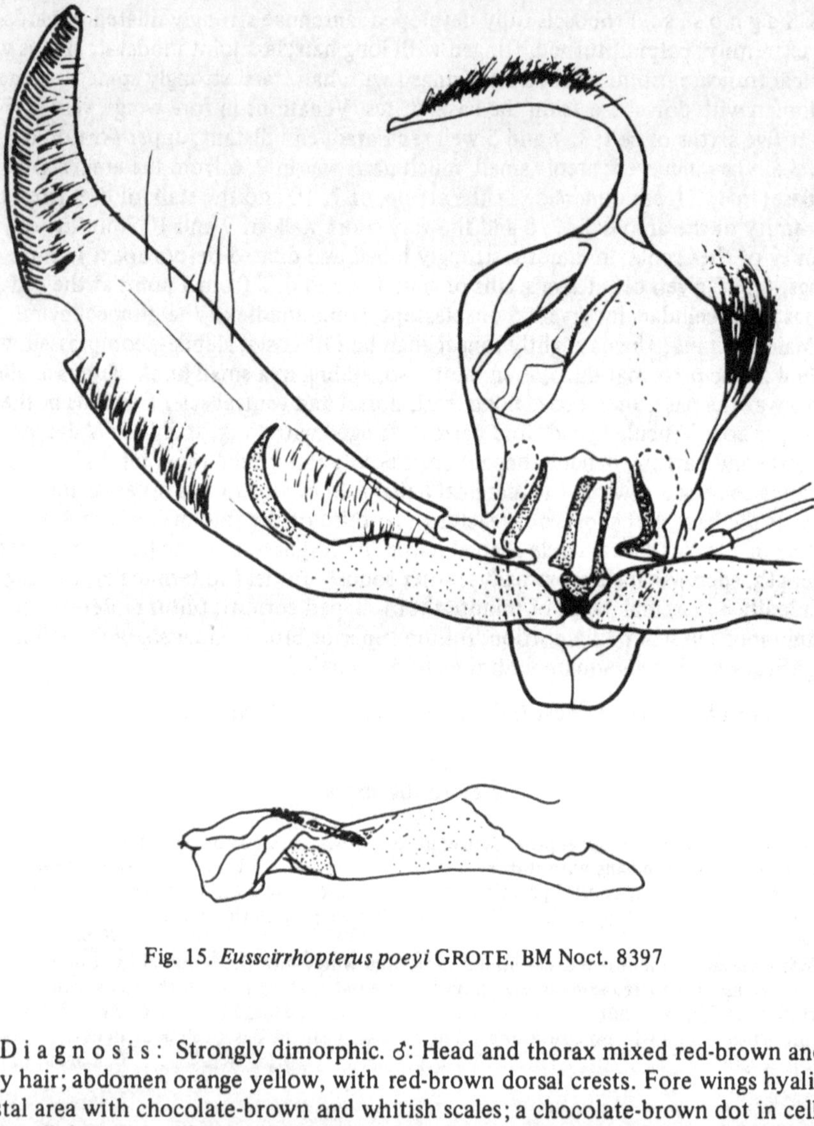

Fig. 15. *Eusscirrhopterus poeyi* GROTE. BM Noct. 8397

D i a g n o s i s : Strongly dimorphic. ♂: Head and thorax mixed red-brown and grey hair; abdomen orange yellow, with red-brown dorsal crests. Fore wings hyaline; costal area with chocolate-brown and whitish scales; a chocolate-brown dot in cell; discocellular mark oblique, olive, edged with chocolate-brown, and with a a silvery lunule at middle; dorsal area to the submedian fold and to near the tornus strongly sprinkled with olive above vein 1 and suffused with grey below it; oblique silvery streaks on olive patches before and just beyond the middle, above vein 1, and continued at an angle to the dorsum as chocolate-brown lines; a chocolate-brown apical patch, and a terminal series of points. Hind wings hyaline; costal area clothed with chocolate-brown and grey scales to just into the cell, leaving a hyaline patch above vein 7 towards the apex; inner area orange yellow; a terminal chocolate-brown dentate band, expanding into a patch with a yellow spot in it towards the anal angle. ♀: Head white, with some chocolate-brown scales on the palpi and on the vertex; antennae white, ringed with brown; thorax clothed with chocolate-brown and grey hair; pectus and legs white, the latter sprinkled with rufous; abdomen orange, with the underside whitish. Fore wings dark chocolate-brown; costal, dorsal and terminal areas sprinkled with white; a silvery-white fascia from the base of the costa, in a below the cell; the minute chocolate-brown cell- and the oblique, white-centred discocellular spot connected with the dark costal area; then the fascia runs obliquely to the costa before the apex; area below it strongly sprinkled with olive; a terminal series of small lunules. Hind wings orange yellow; a terminal chocolate-brown band, leaving some yellow on the inner side of the terminal line from vein 4 to the anal angle; cilia spotted with white and rufous, but yellow at the anal angle. Length of fore wing 21 mm (♂), 24 mm (♀).

D i s t r i b u t i o n : Cuba, Haiti, Jamaica, Panamà, Mexico, Guatemala, Bolivia.

Euscirrhopterus klagesi (JORDAN)

1908 *Euschirropterus klagesi* JORDAN, Novit. zool., **15**:254; pl. 9, fig. 9, ♀. – T e r r a t y p i c a : Venezuela, Ciudad Bolivar (Type, ♀, in British Museum).
1908 *Euschirropterus klagesi* aberr. *fuscus* JORDAN, Novit. Zool., **15**:254; – T e r r a t y - p i c a : Venezuela (Type in British Museum).
1914 *Euschirropterus poeyi pulverosa* DYAR, Proc. U.S. Nation. Mus., **47**:175. – T e r r a t y p i c a : Panamà, La Chorrera (Type, ♀, in U.S. National Museum).
1966 *Euscirrhopterus klagesi* – TODD, Proc. U.S. Nation. Mus., **120**:7; fig. 3, 4.

D i a g n o s i s : ♂: "... it resembles (the male) of *poeyi*, but the hindwing does not possess a dark marginal band, and it has but a single, well-developed, orange terminal spot". (Todd, Proc. U.S. Nat. Mus., 120 :7). ♀ Head white above, rest of the head and the antennae brown; thorax pale grey with some white; abdomen dark brown with orange subdorsal stripes. Fore wings rather broader than in *E. disciferus*, grey brown; a whitish streak, more or less extending to the costa near the base; cell with an oval brown spot; discocellulars with a large brown, lunulate spot; costa marked with brown; distal area brown, narrower than in *E. disciferus*; tornal area more yellowish. Hind wings very pale yellow, with a grey brown terminal band, typically about one-third of the wing in width; a whitish and dark brown subterminal lunule near the anal angle. Length of fore wing 17 mm (♂), 17–21 mm (♀).

D i s t r i b u t i o n : Venezuela, Panamà, ? Ecuador,. ? Brazil.

Euscirrhopterus walkeri (HAMPSON)

1901 *Euschirropterus valkeri* HAMPSON, Cat. Lepidopt. Phal. Brit. Mus., **3**:619–620; fig. 270. – T e r r a t y p i c a : Chile, Valparaiso (Type, ♂, in British Museum).
1966 *Euscirrhopterus walkeri* – TODD, Proc. U.S. Nation. Mus., **120**:8.

D i a g n o s i s : Head and thorax clothed with chocolate-brown and grey hair; head yellowish white above; abdomen orange, with the dorsal crests and the extremity mixed grey and brown. Fore wings chocolate-brown, thickly sprinkled with white; a white fascia from the base in and below the cell, leaving the elongate cell-spot and the discocellular mark connected with the dark costal area, then running obliquely to the costa, before the apex, and with an oblique dark line from its angle to the dorsum; a terminal series of chocolate-brown striae. Hind wings orange yellow, with broad dark-brown terminal band; a yellow spot on the termen above the anal angle. Length of fore wing 16 mm.

Distribution: Chile, Argentina.

Euscirrhopterus walkeri walkeri (HAMPSON)

Diagnosis: Vide that of the species.

Distribution: Chile.

Euscirrhopterus walkeri tucumanus (JORDAN)

1908 *Euschirropterus klagesi tucumanus* JORDAN, Novit. Zool., **15**:254. – Terra typica: Tucuman, Jujuy, Argentina (Type, ♀, in British Museum).
1966 *Euscirrhopterys walkeri tucumanus* TODD, Proc. U.S. Nat. Museum, **120**:9, fig. 1, 2.

Diagnosis: Head and thorax darker, the former with fewer white scales; in ♂, the white areas are much broader than in the typical subspecies; ♀ with the fore wings broader, and the termen less oblique; the white area begins further from the base and is less widened towards the costa before the apex. Length of fore wing 14 mm (♂), 18–23 mm (♀).

Distribution: Argentina.

Euscirrhopterus disciferus (HAMPSON)

1901 *Euschirropterus discifera* HAMPSON, Cat. Lepidopt. Phal. Brit. Mus. **3**:621; pl. 51, fig. 5. – Terra typica: Santa Catarina, Brazil (Type, ♀, in British Museum).
1966 *Euscirrhopterus disciferus* – TODD, Proc. U.S. Nation. Mus., **120**:9.

Diagnosis: ♀: Head pale yellow; palpi, sides of frons, antennae dark brown; antennal shaft sprinkled with white above; thorax whitish grey; collar and legs dark brown; abdomen dark brown, with broad orange subdorsal stripes, connected by dorsal segmental lines, and not reaching the extremity. Fore wings grey-brown, with a large greyish-white semicircular patch on the costa from the base to near the apex, extending to below middle of the cell; costal area sprinkled with brown, and leaving a large dark brown discocellular lunule; the grey-brown area shading into dark brown towards the greyish-white patch. Hind wings grey-brown, with a yellow patch in the basal area, extending from the anal margin to the subcostal vein and to beyond the cell; a yellow and dark brown subterminal lunule in the submedian interspace. Length of fore wing 20–23 mm.

Distribution: Brazil (only ♀♀ from the type locality known).

Genus **Schalifrontia** HAMPSON

1901 *Schalifrontia* HAMPSON, Cat. Lepidopt. Phal. Brit. Mus., 3:654. – Species typica: *Schalifrontia furcifer* HAMPSON (by original designation).

Diagnosis: Proboscis fully developed; antennae slightly dilated towards the extremity; palpi oblique, the second joint fringed with long hair in front, the third small, naked; frons with a three-grooved conical spine ending in a minute fork, the lateral part of frons produced into lateral walls; eyes hairy; tibiae with appressed scales. Nervation: vein 2 from two-thirds of cell; 3, 4, 5 well separated, distance between 4 and 5; discocellulars oblique, incurved; 6 from the upper cell-angle; 7 and the long stalk of 8 + 9 from the extremity of the very narrow areole; 10 from upper margin of areole, near its extremity. Hind wings with vein 2 from slightly beyond half of cell; 3 and 4 separated, discocellulars faintly incurved; 5 obsolescent, from middle of discocellulars; 6, 7 from the upper cell-angle.

Male genitalia: Uncus relatively very small and slender, faintly arched in basal half, then faintly concave and ending in a minute hook. Tubus analis broad, well sclerified both dorsally and ventrally. Vinculum narrow, without hairfringe. Valva very elongate, narrow, banana-shaped; apex produced into a short lobe edged on the proximal side with a few stiff setae; fold of sacculus about half of valva in breadth, reaching slightly beyond one-third of valva, and ending in a long, nearly straight harpe, more than one-third of valva in length, and ending in a hook. Aedeagus about five-eighths of valva in length, very slender, roughly S-shaped, with the proximal portion spoon-shaped, and with a subterminal patch of minute cornuti; fultura inferior with the proximal angles produced into slender processes, and the distal margin angled inwards; median portion membraneous proximally. Saccus rounded with the sides slightly incurved.

Distribution: Brazil.

Schalifrontia furcifer HAMPSON

(fig. 16)

1901 *Schalifrontia furcifer* HAMPSON, Cat. Lepidopt. Phal. Brit. Mus., 3:654; fig. 288. – Terra typica: Brazil, Santa Catarina (Type, ♀, in British Museum).

Diagnosis: Body and wings black; collar with orange-yellow lateral patches; coxae with some yellow hair; abdomen in ♂ uniformy dark brown or black on the underside, with a yellow lateral line on urites IV to VII; in ♀, with narrow lateral and broad ventral yellow stripes, not quite reaching the extremity. Markings on the fore wings yellowish white in ♂, yellow-orange in ♀; a stripe from the base to the origin of vein 2, extending to the submedian fold and into middle of cell; an oblique band beyond cell from below costa to above vein 2 near the termen. Hind wings with a wedge-shaped orange-yellow fascia widening from the base to near the termen, where it extends from just below vein 2 to vein 7, at the base confluent with a fascia on the anal area extending to near the anal angle. Length of fore wing 21 mm.

Distribution: Brazil.

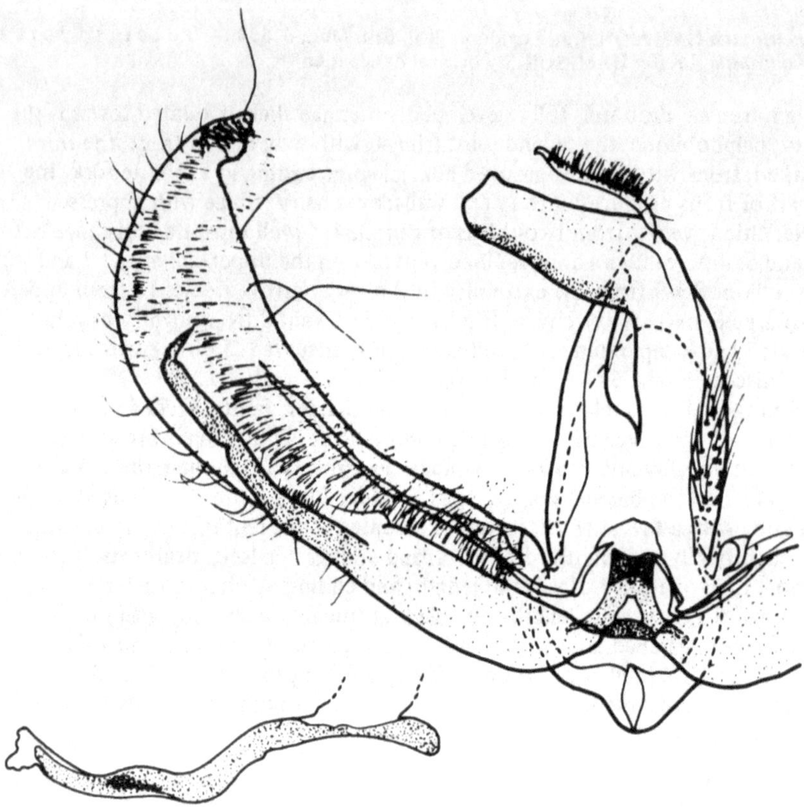

Fig. 16. *Schalifrontia furcifer* HAMPSON. BM Noct. 8403

Genus **Alypia** HÜBNER

1818 *Alypia* HÜBNER, Zutr. exot. Schmetterl.,:22. – Species typica: *Alypia quadriguttalis* HÜBNER = *Sesia octomaculata* FABRICIUS.
1873 *Androloma* GROTE, Bull. Buffalo Soc., 1:30. – Species typica: *Alypia maccullochii* KIRBY.

Diagnosis: Proboscis fully developed; antennae distinctly dilated towards the extremity; palpi upturned, with the second joint fringed with very long hair in front, the third joint moderate and hairy; frons with a truncate conical prominence with raised rim at the extremity; tibiae with rough hair on the outer side; mid- and hindtibiae with long hair; base of abdomen dorsally with rough hair. Fore wings sometimes with the costa dilated; a groove of ribbed membrane below it. Nervation: in fore wings, vein 2 from three-quarters to five-sixths of the cell; 3, 4, 5 well separated, equidistant; discocellulars faintly incurved; 6 from the lower cell-angle; areole narrow; 7 and the short stalk of 8 + 9 from its extremity; 10 from its upper angle. In hind wings, vein 2 from about three-quarters of the cell; 3, 4 and 6, 7 from the cell-angles; discocellulars incurved; 5 obsolescent, from middle of discocellulars.

Male genitalia: Uncus about half of costa in length, narrow, arched, with a small terminal hook. Tubus analis elongate and broad, sclerified both above and below. Vinculum rather broad, with a broad fold which is slightly lobed proximally and bears a large tuft of long stiff hair. Valva elongate, more or less banana-shaped but broadening distally; apex rounded, bearing a few stiff setae; fold of sacculus moderately broad, reaching to about mid-valva, and bearing a slender harpe, about as long as the uncus, straight for the proximal two-thirds, then arched. Aedeagus distinctly longer than the costa, slender proximally, gradually becoming more robust, but narrowing again very much at the distal quarter; terminal section again broad, bearing a slender, almost filiform projection; towards the end of the broad portion an elongate patch of small cornuti; fultura inferior broad, with the sides produced into a small pointed process near their bases; distal margin angled at middle. Saccus rounded.

Distribution: North America; Argentina.

Key to the species

1	Fore wings with the costa dilated, with a groove of ribbed membrane below it	2
1′	Fore wings without dilation or groove	4
2	Fore wings of ♂ with the groove of ribbed membrane extending from the base to beyond middle	*A. maccullochii*, p. 54
2′	Fore wings of ♂ more elongate, and with the ribbed membrane postmedial only	3
3	Fore wings with the markings yellow	*A. disparata*, p. 56
3′	Fore wings with the markings white	*A. brannani*, p. 56
4	Fore wings with silvery-blue markings	5
4′	Fore wings without silvery-blue markings	8
5	Hind wings with the markings white	6
5′	Hind wings with the markings yellow	*A. ridingsi*, p. 52
6	Base of hind wings black	7
6′	Base of hind wings white, without any black	*A. matuta*, p. 51
7	Markings of fore wings pale yellow	*A. octomaculata*, p. 49
7′	Markings of fore wings light buff, very large	*A. wittfeldii*, p. 51
8	Hind wings with a subbasal yellow spot	*A. dipsaci*, p. 52
8′	Hind wings without subbasal yellow spot	*A. langtoni*, p. 52
9	Hind wings with a yellow spot at the cell-end, a patch beyond the cell between veins 2 and 7, and with the veins black	*A. mariposa*, p. 53
9′	Hind wings with the markings otherwise placed	
10	Hind wings black, unmarked	
10′	Hind wings orange yellow, with dark terminal band	
11	Fore wings with an oblique pale yellow band and without other markings	*A. lulesa*, p. 54
11′	Fore wings with basal and cell-markings besides a pale orange postmedial band	*A. pampeana*, p. 54
12	Hind wings with a black discocellular mark	*A. australis*, p. 53
12′	Hind wings without black discocellular mark	*A. ornata*, p. 54

Alypia octomaculata (Fabricius)

(fig. 17)

1775 *Sesia octomaculata* FABRICIUS, Syst. Entomol.,:830. – Terra typica: North America (Type not designated).

1782 *Noctua albomaculata* STOLL, Uitheemsche Kapellen, 4: pl. 345 C. – Terra typica: North America (Type not designated).

1791 *Zygaena bimaculata* GMELIN, Syst. Nat., 1 (5):2398. – Terra typica: North America (Type not designated).

1797 *Sphinx octomaculata* – SMITH & ABBOTT, Lepidopt. Georgia, 1: pl. 44.
1818 *Alypia quadriguttalis* HÜBNER, Zutr. exot. Schmetterl., 1:22; Nr. 3363. – Terra typica: North America.
1825 *Agarista octomaculata* – LATREILLE, Encyclop. method., 9:803.
1854 *Euchromia bimaculata* – WALKER, List Lepidopt. Het. Brit. Mus., 1:275.

Diagnosis: Body and wings black; antennae annulated with white above; palpi yellowish at the base and extremity of the second joint; frons yellowish at the sides; collar with a yellow medial spot; tegulae pale yellow; fore and mid-tibiae orange above; abdomen with a whitish dot at the base, and a more or less developed whitish or yellowish dorsal stripe on terminal segments, absent in ♂. Fore wings with a more or less squarish, triangular and rounded pale yellow medial spot, from the subcostal vein to the submedial fold; an elliptical spot beyond the cell from vein 7 to below vein 3; silvery-blue streaks from the base below the costa and above vein 1, a discocellular lunule, an oblique postmedial bar between veins 1 and 2, and another between veins 6 and 8. Hind wings with a white patch from near base which remains black, to middle, and from the subcostal vein or below it to vein 1, more rounded in ♀, and sometimes in ♂ as well (= *albomaculata*). Length of fore wing 14–18 mm.

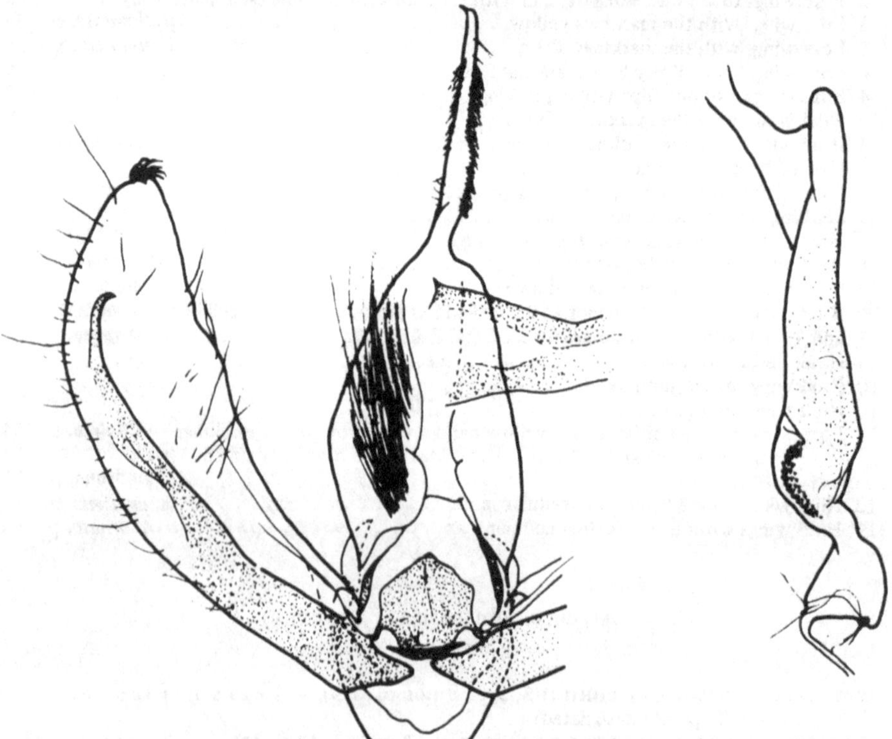

Fig. 17. *Alypia octomaculata* (FABRICIUS). BM Noct. 8398

E a r l y s t a g e s : Larva: young, whitish tinged with orange-red on middle of each segment, with numerous irregular transverse brownish lines; strongly developed dorsal tubercles, especially on the 4th, 5th, 6th and 11th segments, acuminate, brownish-black; head and dorsum from the first to the twelfth segments pinkish orange; eleventh segment with irregular lateral white spot. Full-grown: head and body segments 1st to 12th, orange dotted with black; second and third segments with transversal series of black dots; other segments with orange band dotted with black, and a long whitish hair; between each band six black lines and dots connected by interrupted black lines; legs orange at the base, black externally and at the extremity, except the anal pair; tenth and eleventh segments with a large lateral white spot; fourth to ninth segments also with lateral white spots; orange bands dorsally interrupted on seventh to ninth segments. Pupa underground, in an earthy cocoon with a gummy secretion, full fed in July.

D i s t r i b u t i o n : Greatest part of the United States and Canada.

Alypia wittfeldii HY. EDWARDS

1883 *Alypia wittfeldii* HY. EDWARDS, Papilio, **3**:34. – T e r r a t y p i c a : U.S.A., Florida, Indian River (Type in Amer. Museum Nat. Hist.).
1901 *Alypia octomaculata* -- HAMPSON, Cat. Lepidopt. Phal. Brit. Mus., **3**:637.

D i a g n o s i s : Similar to *A. octomaculata.* Markings of fore wings light buff; proximal spot more elongate, reaching nearly across the wing from the subcostal vein to close to the dorsum; distal spot pyriform, much larger than in *A. octomaculata.* In the hind wings, spots pure white, small, the proximal spot oblong ovate, the distal spot ovate; base and anal margin broadly black. Size as in *A. octomaculata.*

D i s t r i b u t i o n : U.S.A.

Alypia matuta HY. EDWARDS

1883 *Alypia matuta* HY. EDWARDS, Papilio, **3**:33. – T e r r a t y p i c a : U.S.A. (Type in Amer. Museum Nat. Hist.).
1901 *Alypia octomaculata* – HAMPSON, Cat. Lepidopt. Phal. Brit. Mus., **3**:637.

D i a g n o s i s : Very similar to *A. octomaculata.* Spots on the fore wings much larger; basal patch sub-triangular, straight along the dorsum, more or less rounded costally, ending based in a fine point on the median vein; distal patch ovate, irregular in outline costally and somewhat constricted in the middle. Hind wings with the basal third pure white, without any black; distal spot much larger and more pyriform than in *A. octomaculata.* Size as in the latter species.

E a r l y s t a g e s : Larva: wholly bright lemon yellow, with waved transverse lines of black, and with black spines.

D i s t r i b u t i o n : U.S.A. (Colorado, Missouri, Alabama).

Alypia dipsaci GROTE & ROBINSON

1868 *Alypia dipsaci* GROTE & ROBINSON, Trans. Amer. entomol. Soc., **1**:326; pl. 6, fig. 27. – Terra typica: U.S.A., California (Type in Amer. Museum Nat. Hist.).
1869 *Agarista dipsaci* – BOISDUVAL, Ann. Soc. entomol. Belg., **12**:68.

Diagnosis: Body and wings black; palpi with a yellowish lateral mark; tegulae with a yellowish stripe; fore and mid-tibiae orange. Fore wings with an almost medial, rather irregular spot from the subcostal vein to the submedian fold; an elliptical patch beyond the cell between veins 2 and 8; silvery-blue streaks from the base below the costa and above vein 1. Hind wings with the antemedial yellow spot from the subcostal vein to the submedian fold, and a patch beyond the cell from vein 7 to below vein 3. Length of fore wing 14 mm.

Distribution: U.S.A.

Alypia langtoni COUPER

1865 *Alypia langtoni* COUPER, Canad. Naturalist. (2) **2**:64, 460; fig. – Terra typica: Canada (Type in Mc Gill Museum, Montreal).
1868 *Alypia sacramenti* GROTE & ROBINSON, Trans. Amer. entomol. Soc., **1**:327; pl. 6, fig. 38. – Terra typica: U.S.A., California (Type in Amer. Museum Nat. Hist.).
1884 *Alypia hudsonica* HY. EDWARDS, Papilio, **4**:43. – Terra typica: Hudson's Bay District (Type in Amer. Museum Nat. Hist.).

Diagnosis: Body and wings black; the second joint of palpi yellowish at the base and at the extremity; sides of frons and a postocular line yellow; collar with a yellow medial spot; tegulae yellow; fore and mid-tibiae orange. Fore wings with a rounded, elliptical or triangular, almost medial spot from the subcostal vein to the submedian fold; an elliptical patch beyond the cell between veins 3 and 8, varying in shape; a silvery-blue spot below the costa near the base, a streak above vein 1, and a discocellular bar. Hind wings with a variable spot beyond the cell between veins 3 and 7. Length of fore wing 15–16 mm.

Distribution: Most of the United States and Canada.

Alypia ridingsi GROTE

1864 *Alypia ridingsi* GROTE, Proc. entomol. Soc. Philadelphia, **3**:521; pl. 5, fig. 1. – Terra typica: U.S.A., Colorado (Type in U.S. National Museum).
1892 *Androloma ridingsi* – KIRBY, Syn. Cat. Lepidopt. Heterocera,:34.

Diagnosis: Body and wings black; collar with small yellow spots; mid-tibiae orange. Fore wings with a triangular antemedial spot from the subcostal vein to the submedian fold; a spot in the upper cell-end; a curved band beyond the cell between veins 2 and 7, intersected by the black veins; cilia sometimes white at the apex. Hind wings with a yellow spot in the cell-end, and a patch beyond cell between veins 2 and 7, crossed by the black veins. Length of fore wing 14–15 mm.

Early stages: Larva: violaceous-black, browner at the segmental junctures; each segment with a large dorsal silvery-white square spot marked with violaceous brown, and a lateral longitudinal band spotted with black, and with yellow towards the edges of the segment; the first segment with a dorsal orange bar; underside pale brown, sometimes whitish, spotted with brown. Legs black; prolegs violaceous brown, tipped with black. A number of moderately long white setae on each segment, arising from black papillae placed on the black bands and spots. Head black, spotted with yellow, with a few short white setae. Oenothera bistorta, Oe. dentata, Clarkia rhomboidea, Eulobus californicus.

Distribution: The Western United States.

Alypia mariposa GROTE & ROBINSON

1868 *Alypia mariposa* GROTE & ROBINSON, Trans. Amer. entomol. Soc., **1**:329; pl. 6, fig. 40. – Terra typica: U.S.A., California (Type in Amer. Museum Nat. Hist.).
1869 *Agarista mariposa* – BOISDUVAL, Ann. Soc. entomol. Belg., **12**:70.
1872 *Alypia lunata* STRETCH, Zygaenae and Bombyces of North America,:15; pl. 1, fig. 6. – Terra typica: U.S.A., California (Type in Calif. Acad. Sci., San Francisco).

Diagnosis: Body and wings black; collar spotted with yellow; mid-tibiae orange. Fore wings with some yellowish sprinkling along vein 1; a subbasal yellowish wedge-shaped spot from just above the subcostal vein to the submedian fold, in which it extends to the middle of the wing; an elliptical spot at the cell-end; postmedial spots between veins 2 and 3, and 6 and 8, sometimes conjoined by a curved band (**lunata**); cilia white at the apex. Hind wings with or without a spot in the cell-end; two postmedial spots between veins 1 and 3, and 6 and 8, sometimes broadened to a band extending from vein 1 to vein 8 (**lunata**); cilia white at the apex. Length of fore wing 17 mm.

Early stages: Larva: young yellow, later reddish white, in first stages striped, later spotted with dark, rather irregularly; last segments thickly marked with buff and white. On Clarkea elegans and other plants. Pupa reddish brown; the abdomen with seven raised round glossy white spots.

Distribution: U.S.A. (California).

Alypia australis SCHAUS

1921 *Alypia australis* SCHAUS, Proc. U.S. Nation. Mus., **57**:122. – Terra typica: Argentina, La Farga (Type in U.S. National Museum).

Diagnosis: Body black; some fine white scaling on the neck; a white spot on the shoulders. Fore wings fuscous brown; a cream coloured space from the base to the middle of the wing, its anterior edge barely entering the cell at the base, then expanding to near the subcostal vein, curved down across the median vein and the base of vein 2, the lower edge from just below the submedian vein at the base, then upcurved toward the cell; a subterminal cream-coloured spot from vein 2 to vein 8, cut by the veins mostly finely black, vein 5 being more heavily marked. Hind wings orange-yellow, the termen broadly black, especially from vein 2 to the anal margin; a black discocellular streak. Length of fore wing 14 mm.

Distribution: Argentina.

Alypia ornata JÖRGENSON

1935 *Alypia ornata* JÖRGENSON, An. Mus. Argent. Cienc. Nat., 38 : 124; pl. 3, fig. 31. – Terra typica: Argentina, Cazada and La Junta (Type in Museo Argentino de Ciencia Nat.).

Diagnosis: Very similar to *A. australis* SCHAUS. Deep black; collar finely edged with yellow. Fore wings with an orange spot on the tegulae; a broad ochre- and orange-coloured basal stripe, running below the median vein to the middle of vein 5, not entering the cell; a large triangular ochre-coloured cell-spot, with the top toward the base of the wing; postmedial band broad, ochreous, divided into spots by the black veins, and interrupted between veins 5 and 6, with only a few yellow atoms in interspace v. Hind wings orange-yellow, with a broad black terminal band; base finely black; there is no black discocellular mark. Length of fore wing 15 mm.

Distribution: Argentina.

Alypia pampeana JÖRGENSON

1935 *Alypia pampeana* JÖRGENSON, An. Mus. Argent. Cienc. Nat., 38:124–125; pl. 3, fig. 30. – Terra typica: Argentina, Catamarca, Esquina Grande (Type in Museo Argentino de Ciencia Nat.).

Diagnosis: Rather similar to *A. ornata*. Black; tegulae with a large orange spot. Fore wings as in *A. ornata*, but with the markings pale orange; the basal streak darker orange at the base, beginning very broadly, but narrowing soon to form an axe-shaped terminal spot (in *A. ornata* the extremity is cut obliquely, and only reaches vein 5); cell-spot as in *A. ornata*, but larger; postmedial spot narrower, unevenly divided into five spots by the black veins; the last spot very small; the spot is moreover broadly broken between veins 5 and 6. Hind wings solid black, faintly glossed with blue. Length of fore wing 15 mm.

Distribution: Argentina (prov. Catamarca).

Alypia lulesa KÖHLER

1940 *Alypia lulesa* KÖHLER, Rev. Soc. entomol. Argent., 10:369. – Terra typica: Argentina, Tucumàn, Quèbrada de Lules (Type in Museo Argentino de Ciencia Nat.).

Diagnosis: Blackish grey; wings with the veins distinctly outlined; fore wings with a dark blue sheen; an oblique straight pale yellow band from above the discocellulars to the tornus, not quite reaching the costa; its breadth about one-quarter of its length. Length of fore wing 14 mm.

Distribution: Argentina (prov. Tucumàn).

Alypia maccullochii (KIRBY)

1837 *Alypia maccullochii* KIRBY, Fauna Bor.-Amer., 4:301, pl. 4, fig. 5. – Terra typica: Nova scotia seu Canada (Type not designated).
1873 *Androloma maccullochii* – GROTE, Bull. Buffalo Soc., 1:30.
1901 *Alypia (Androloma) maccullochi* – HAMPSON, Cat. Lepidopt. Phal. Brit. Mus., 3:635–636; fig. 279.

Diagnosis: Body and wings black; basal half of the antennae ringed with white; a postocular streak, slight spots on the collar and stripes on the tegulae pale yellow; fore and mid-tibiae orange, the former with an black spot; tarsi ringed with white; abdomen with slight yellow dorsal spots at the base. Fore wings with a triangular yellow patch from near the base to beyond the middle, and from the subcostal vein to the submedian fold, intersected by the black median vein; a spot at the upper cell-end; a band beyond the cell between veins 2 and 8, intersected by the black veins; a silvery-blue streak below the basal half of the costa, and on vein 1, a V-shaped mark enclosing the spot in the cell-end, and a curved subterminal band. Hind wings with a white or pale orange patch in the cell conjoined to a spot below its extremity; a postmedial band between veins 2 and 7 intersected by the black veins. Length of fore wing 13–15 mm.

Distribution: The greater part of the United States and Canada: Mexico.

Alypia maccullochii maccullochii (KIRBY)

Diagnosis: Vide that of the species.

Distribution: Most of the area, except California.

Alypia maccullochii lorquini (GROTE & ROBINSON)

1868 *Alypia lorquini* GROTE & ROBINSON, Trans. Amer. ento. Soc., **1**:328, t. 6 fig. 39. – Terra typica: California (Type in Amer. Mus. Nat. Hist).
1869 *Agarista lorquini* – BOISDUVAL, Ann. Soc. ent. Belg., **12**:69.
1892 *Androloma lorquini* – KIRBY, Syn. Cat. Lep. Het., p. 34.
1938 *Androloma maccullochii lorquini* MCDUNNOUGH, Check List Lep. Can. & U.S.A., **1**:53. –

Diagnosis: Fore wings with the postmedial band elliptical.

Distribution: California.

Alypia maccullochii similis (STRETCH)

1872 *Alypia similis* STRETCH, Zyg. & Bomb. N. Amer., p. 14, t. 1 fig. 5. – Terra typica: Mexico (Type in Amer. Mus. Nat. Hist.).
1892 *Androloma similis* – KIRBY, Syn. Cat. Lep. Het., p. 34
1938 *Androloma maccullochii similis* – MCDUNNOUGH, Check List Lep. Can. & U.S.A., **1**:53.
1874 *Agarista edwardsi* BOISDUVAL, Rev. Mag. Zool., (3) **2**:63. – Terra typica: North America (Type in Paris Museum [?]).
1883 *Alypia similis*, var. *conjuncta* HY. EDWARDS, Papilio, **3**:38.

Diagnosis: Fore wings with the markings partly confluent; hind wings with the markings pale yellow.

Distribution: Mexico, California.

Alypia disparata Hy. Edwards

1884 *Alypia disparata* HY. EDWARDS, Papilio, **4**:13. – Terra typica: Mexico (Type in Amer. Mus. Nat. Hist.).
1887 *Alypia gracilenta* GRAEF, Entomologica americana, **3**:41. – Terra typica: Texas (Type in U.S. National Museum)
1892 *Androloma desperata* KIRBY (ex errore), Syn. Cat. Lepidopt. Heterocera,:35.
1938 *Androloma disparata* – MCDUNNOUGH, Check List Lepidopt. Canacta and U.S.A., **1**:54.

Diagnosis: Body and wings black; postocular stripe and sides of frons yellowish; collar spotted with yellow; tegulae striped with yellow; fore and mid-tibiae orange. Fore wings with an elliptical medial yellow spot from the subcostal vein to the submedian fold; an elliptical yellow spot beyond the cell between veins 3 and 7; a silvery-blue streak on the base of vein 1, a discocellular bar, and a postmedial bar between the submedian fold and vein 1. Hind wings with a purplish gloss; a subbasal white patch between the subcostal vein and vein 1, and a spot beyond the cell between veins 4 and 6. Length of fore wing 17–18 mm.

Distribution: Mexico, U.S.A. (Texas).

Alypia brannani Stretch

1872 *Alypia brannani* STRETCH, Zygaenae & Bombyces of North America, :8; pl. 1, fig. 8. – Terra typica: California (Type in Calif. Acad. Sci., San Francisco).
1892 *Androloma brannani* – KIRBY, Syn. Cat. Lepidopt. Heterocera,:34.

Diagnosis: Body and wings black; basal palpal joint with yellow hair; collar pale yellow; fore and mid-tibiae with tufts of orange hair; abdomen with an indistinct whitish dorsal stripe. Fore wings with an elliptical white spot in and below the middle of the cell, and a similar spot beyond cell between veins 3 and 6. Hind wings with a white spot in and below the cell, and a spot beyond the cell between veins 4 and 6. Length of fore wing 15 mm.

Distribution: Mexico (South California).

Genus Alypioides Grote

1883 *Alypioides* GROTE, Trans. Kansas Acad. Sci., **8**:46. – Species typica: *Alypioides flavilinguis* GROTE = *Agarista bimaculata* HERRICH-SCHÄFFER.
1880 *Pseudalypia* GROTE, Trans. Kansas Acad. Sci., **1**:34 (nec Hy. Edwards 1874). – Species typica: *Eusemia crescens* WALKER = *Agarista bimaculata* HERRICH-SCHÄFER.
1901 *Alypiodes* HAMPSON (ex errore), Cat. Lepidopt. Phal. Brit. Mus., **3**:641.

Diagnosis: Proboscis fully developed; antennae hardly dilated beyond the middle; palpi upturned, with the second joint sparsely fringed with rough hair in front, and the third joint short, roughly scaled; frons with a long conical truncate prominence, with raised rim at the extremity; tibiae clothed with appressed scales; abdomen smoothly scaled. Nervation: in fore wings, vein 2 from four-fifths of cell; 3, 4, 5 separated, equidistant; discocellulars faintly incurved, 6 from the upper cell-angle; areole very narrow; 7 and the long stalk of 8 + 9 from the extremity of the areole; 10 from the upperside of the areole, near the extremity. In hind wings, vein 2 from about three-quarters of cell; 3, 4 and 6, 7 from the cell-angles; 5 obsolescent, from middle of discocellulars, which are incurved.

Male genitalia: Uncus about half of costa in length, slender, faintly arched, with a terminal beaklike hook. Tubus analis broad and elongate. Vinculum rather broad, with the fold moderately broad, bearing a large bunch of long stiff hair. Valva banana-shaped; terminal portion without marginal spinule-like setae; fold of sacculus moderately broad, reaching nearly to two-thirds of valva, ending in a shortish, well arched harpe. Aedeagus about two-thirds of costa in length, moderately robust, faintly arched in the proximal half, straight distally, and deeply incised terminally; fultura inferior cup-shaped. Saccus rounded.

Distribution: U.S.A., Mexico, Guatemala.

Key to the species

1 Collar spotted with yellowish 2
1′ Collar not spotted with yellowish; shoulders with tufts of orange hair . . . *A. walkeri*, p. 59
2 Fore wings with two pale yellow medial patches . . . . *A. geronimo*, p. 59
2′ Fore wings without large medial patches 3
3 Fore wings with a white spot in base of cell . . . . *A. bimaculata*, p. 57
3′ Fore wings without white spot in base of cell; basal area with pale streaks . *A. radians*, p. 58

Alypioides bimaculata (HERRICH-SCHÄFFER)

1850 *Agarista bimaculata* HERRICH-SCHÄFFER, Außereurop. Schmetterl.,: fig. 26. – Terra typica: Mexico (Type in Berlin Museum).
1856 *Eusemia crescens* WALKER, List Lepidopt. Het. Brit. Mus., 7:1774. – Terra typica: Mexico (Type, ♂, in British Museum).
1869 *Agarista grotei* BOISDUVAL, Ann Soc. entomol. Belg., 12:70. – Terra typica: California (Type in British Museum).
1874 *Alypia grotei* – BOISDUVAL, Rev. Mag. Zool., (3) 2:64.
1874 *Alypia trimaculata* BOISDUVAL, Rev. Mag. Zool., (3) 2:65. (Unjustified emendation of *A. bimaculata* HERRICH-SCHÄFFER).
1875 *Pseudalypia crescens* – GROTE, Trans. Kansas Acad. Sci., 1:34.
1883 *Alypioides flavilinguis* GROTE, Trans. Kansas Acad. Sci., 8:46. – Terra typica: Northwest Mexico (Type in U.S. National Museum)
1892 *Alypioides bimaculata* – KIRBY, Syn. Cat. Lepidopt. Heterocera,: 34.
1892 *Alypioides crescens* – KIRBY, Syn. Cat. Lepidopt. Heterocera,:34.
1895 *Alypiodes bimaculata* – COCKERELL, Entomol. News Philadelphia, 6:201.
1896 *Alypiodes bimaculata* var. *dugesi* COCKERELL, Entomol. News Philadelphia, 6:201. – Terra typica: Mexico.

Diagnosis: Body and wings black with bluish gloss; proboscis yellow; a yellow postocular streak; collar and extremities of tegulae with yellowish spots. Fore wings with a yellowish white basal spot; an antemedial wedge-shaped spot in cell; a square spot in cell-end; an oblique spot beyond cell between veins 3 and 7, very variable in shape; a silvery-blue streak below base of costa; a spot above vein 1; a spot in middle of cell; a discocellular bar; cilia usually white apically. Hind wings mostly unspotted; sometimes a yellowish spot beyond cell between veins 4 and 6 (*crescens*), or, besides that, a small yellowish spot below base of vein 2 (*dugesii*).

Distribution: U.S.A. (California, New Mexico), Mex

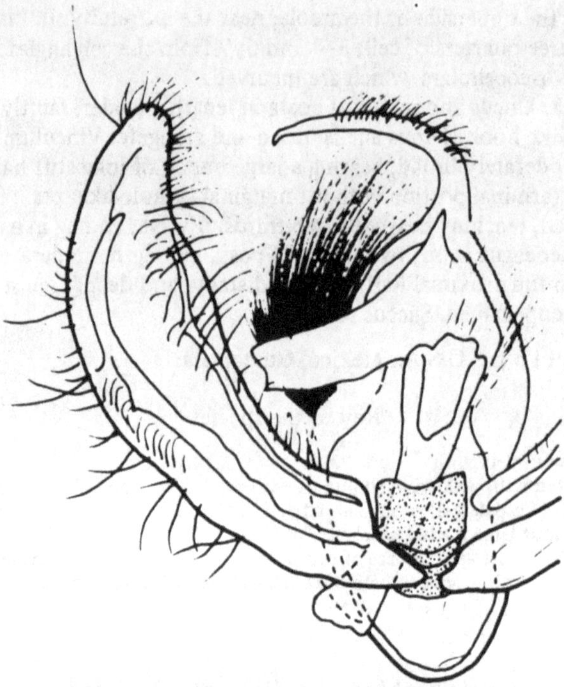

Fig. 18. *Alypioides bimaculata* (HERRICH–SCHÄFFER). No. 48. Leiden Museum

Alypioides radians (FELDER)

1874 *Agarista radians* FELDER, Reise Novara,:pl. 107, fig. 9. – Terra typica: "Moluques", recte Mexico (Type in British Museum).
1874 *Eusemia radians* – BOISDUVAL, Rev. Mag. Zool., (3) **2**:105.
1892 *Phalaenoides* (?) *radians* – KIRBY, Syn. Cat. Lepidopt. Heterocera,:24.
1901 *Alypiodes radians* – HAMPSON, Cat. Lepidopt. Phal. Brit. Mus., **3**:642.

Diagnosis: Body and wings bluish black; proboscis fulvous; first and second joints of palpi, upperpart of frons laterally, and vertex spotted with yellowish white; a yellowish postocular streak; collar, tegulae and metathorax with yellowish spots. Fore wings with yellowish streaks from base to middle below costa, on median vein, in submedian fold, and on vein 1; a metallic blue spot in middle of cell; a discocellular bar with some diffused yellowish before the former, and a square spot between them; an oblique band beyond cell between veins 3 and 7, emitting nervural streaks towards termen, and with a streak below it on vein 2; cilia yellowish apically. Hind wings unspotted; cilia yellowish white. Length of fore wing 22 mm.

Distribution: Mexico.

Alypioides walkeri (DRUCE)

1888 *Pseudalypia walkeri* DRUCE, Ann. Mag. Nat. Hist., (6) **2**:239. – Terra typica: Mexico, Acapulco (Type, ♀, in British Museum).
1889 *Pseudalypia stuartii* SCHAUS, Entomologica americana, **5**:87. – Terra typica: Mexico, Paso de San Juan (Type in U.S. National Museum).
1901 *Alypiodes valkeri* – HAMPSON, Cat. Lepidopt. Phal. Brit. Mus., **3**:642.

Diagnosis: Body and wings black, slightly shot with blue; proboscis fulvous; shoulders with tufts of orange hair. Fore wings with a very oblique postmedial yellowish white band which extends from below costa beyond middle to above tornus, where it is recurved; slight streaks of metallic blue scales at base below costa and above vein 1; a spot in cell-end; a discocellular line; streaks on veins 1, 2, 3 and in the submedian fold from middle to the postmedial band. Hind wings unspotted. Length of fore wing 18–20 mm.

Distribution: Mexico; Guatemala.

Alypioides geronimo (BARNES)

1900 *Pseudalypia geronimo* BARNES, Canad. Entomologist, **32**:44. – Terra typica: U.S.A., Arizona, Huachuca Mounts (Types, ♂, ♀, in U.S. National Museum).
1938 *Alypioides geronimo* – MC DUNNOUGH, Check List Lepidopt. Canada and U.S.A., **1**:54.

Diagnosis: Body black with a slight metallic gloss; proboscis yellow; leg pilosity orange on inner side; flanks with a tuft of orange hair at the base of wings. Fore wings brownish-black; two large square pale yellow medial patches forming a band; the costal patch edged by metallic blue scales on inner side; veins partly streaked with metallic blue. Hind wings slightly browner black, unspotted. Length of fore wing 17 mm.

Distribution: U.S.A. (Arizona).

Genus Psychomorpha HARRIS

1839 *Psychomorpha* HARRIS, Amer. J. Sci., **36**:319. – Species typica: *Psychomorpha glaucopis* HARRIS = *Noctua epimenis* DRURY.

Diagnosis: Proboscis fully developed. Antennae bipectinated, with moderate branches to the apex. Eyes small, elliptical. Palpi porrect, short, clothed with long hair. Frons with a large pointed, conical prominence. Thorax, abdomen and tibiae with long, rough pilosity; thorax without crests. Nervation: in fore wings, vein 2 from about two-thirds of cell; 3, 4, 5 approximated, 4 and 5 slightly more apart than 3 and 4; discocellulars incurved; 6 from the upper cell-angle; areole very narrow; 7 and the very short stalk of 8 + 9 from the extremity of areole; 10 from the upper angle thereof. In hind wings, vein 2 from very slightly beyond middle of cell; 3, 4 and 6, 7 from the cell-angles; discocellulars faintly incurved; 5 obsolescent, from middle of discocellulars.

Male genitalia: Uncus relatively short, narrow, lanceolate, strongly elbowed before middle. Tubus analis short, broadening and sclerified terminally. Fold of vinculum narrow, hairy, but without a distinct hair-bunch. Valva banana-shaped; fold of sacculus half as broad as the valva, reaching to slightly farther than one-third of the valva, broadening terminally, and ending in a moderate sickle-shaped harpe. Aedeagus about as long as the costa, rather slender, very faintly bent before one-third, and again, but in the other direction, after middle; terminal portion broadening and not sclerified; fultura inferior small, squarish, with the middle of the proximal margin and the proximal angles produced and pointed; fultura superior bearing lateral hook-shaped lobes. Saccus broad, vaguely squarish.

Distribution: U.S.A.

Key to the species

1 Hind wings with a cupreous-red patch, extending to termen . . . . *P. euryrhoda*, p. 61
1' Hind wings with an orange-red patch, leaving a black terminal band . . . *P. epimenis*, p. 60

Psychomorpha epimenis (DRURY)

1780 *Noctua epimenis* DRURY, Illustr. Exot. Entomol., 3:40; pl. 29, fig. 2. – Terra typica: North America (Type not designated).
1839 *Psychomorpha glaucopis* HARRIS, Amer. J. Sci., 36:319. – Terra typica: North America.
1892 *Psychomorpha epimenis* – KIRBY, Syn. Cat. Lepidopt. Heterocera,:36.

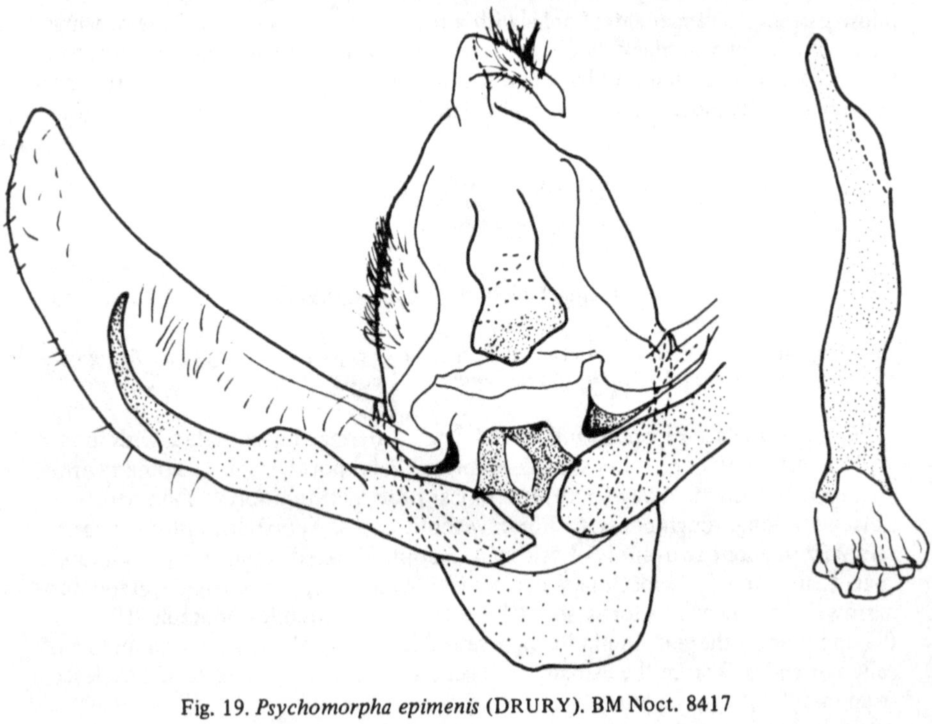

Fig. 19. *Psychomorpha epimenis* (DRURY). BM Noct. 8417

D i a g n o s i s : Body blackish-brown, with a few greyish hairs. Fore wings blackish-brown, slightly tinged with coppery red and sprinkled with a few white and metallic blue scales on basal half and termen; a large white postmedial patch from costa to the submedian fold; a diffused curved silvery-blue subterminal line. Hind wings blackish-brown, with a broad orange-red postmedial band from below costa to above tornus; cilia with a silvery-blue gloss. Length of fore wing 11–13 mm.

E a r l y s t a g e s : Larva: Head red-brown with four black spots on each side; body slightly enlarged on the 12th segment; each segment with four purple-brown transverse lines alternating with four pale yellow ones; a reddish dorsal patch on the 12th segment, containing four tubercles in a square; cervical shield and anal plate red, spotted with black. Living on Vitis.

D i s t r i b u t i o n : U.S.A.

Psychomorpha euryrhoda HAMPSON

1910 *Psychomorpha euryrhoda* HAMPSON, Cat. Lepidopt. Phal. Brit. Mus., 9:425; pl. 146, fig. 7. – T e r r a t y p i c a : U.S.A., Florida (Type, ♂, in British Museum).

D i a g n o s i s : Body blackish-brown with some metallic blue scales. Fore wings suffused with cupreous-red and sprinkled with some silvery-blue scales; antemedial line indistinct, formed by diffused silvery-blue scales, oblique from costa to vein 1; a patch of silvery-blue scales above the middle on vein 1; a round cell-spot, indistinctly edged with diffused silvery-blue scales; a broad, somewhat oblique postmedial white band from costa to vein 1; a diffused curved, slightly waved subterminal silvery-blue line; a terminal row of slight patches of blue-white scales. Hind wings fiery carmine-red; basal area dark brown, extending on the anal margin to near tornus; cilia dark brown. Length of fore wing 12 mm.

D i s t r i b u t i o n : U.S.A.

Genus **Phasidia** HAMPSON

1901 *Phasidia* HAMPSON, Cat. Lepidopt. Phal. Brit. Mus., 3:659. – S p e c i e s t y p i c a : *Phasis contraria* WALKER (by original designation).

D i a g n o s i s : Proboscis fully developed; antennae strongly dilated towards the extermity, almost knoblike; palpi obliquely upturned, much longer than in other genera, especially the third joint which is naked and porrect; frontal prominence long, truncate, conical, with raised rims at the extremity; tibiae slightly fringed with hair above. Abdomen with appressed scales. Nervation: in fore wings, vein 2 from seven tenths of cell; 3, 4, 5 well separated, equidistant; discocellulars well incurved; 6 from just below the upper cell-angle; areole minute; 7, 10, 8 + 9 moderately stalked, from the extremity of the areole. In hind wings, vein 2 from slightly beyond five-sevenths of cell; 3, 4 separated; discocellulars incurved; 5 obsolescent, from middle of discocellulars; 6, 7 from the upper cell-angle.

Male genitalia: Uncus small, narrow, nearly straight, with the extremity subacute. Tubus analis short, broadening distally, sclerified ventrally. Vincula narrow; fold of vinculum as broad as the rest of the vinculum, with sparse pilosity. Valva elongate, rather narrow, very faintly broadening terminally, otherwise with costa and sacculus parallel, the former faintly incurved; fold of sacculus one-third narrower than the valva, reaching fully to the middle, and ending in a rather broad, straight harpe, curved only terminally, with the extremity sharp and preceded costally by a small tooth; apex of valva bearing a dense bunch of longish hair; long pilosity also on middle of the valvula. Aedeagus small, about one-half of costa in length, rather slender, faintly boomerang-shaped, with a subterminal row of minute denticulations; fultura inferior small, with the proximal angles produced; sides convex; terminal margin faintly produced, blunt; fultura superior with the sides produced to form slender processes with a valval insertion. Saccus short, rounded.

Distribution: South America.

Phasidia contraria (WALKER)

(fig. 20)

1864 *Phasis contraria* WALKER, List Lepidopt. Het. Brit. Mus., **31**:136. – Terra typica: Not designated (Type ♀, in British Museum).
1892 *Othria meridionalis* SCHAUS, Proc. Zool. Soc. London, **1892**:272. – Terra typica: Brazil, Pebas, Petropolis (Type, ♀, in U.S. National Museum).
1901 *Phasidia contraria* – HAMPSON, Cat. Lepidopt. Phal. Brit. Mus., **3**:659–660; fig. 292.

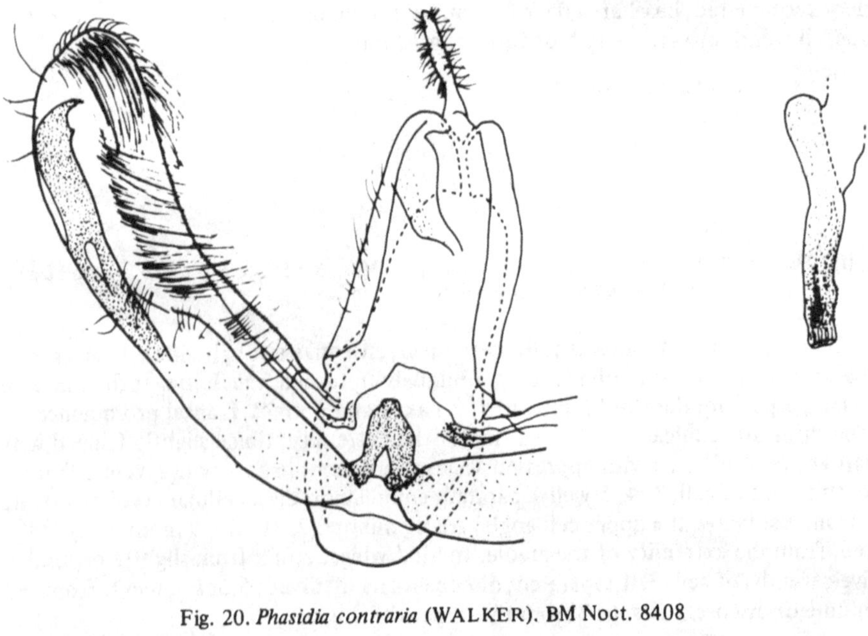

Fig. 20. *Phasidia contraria* (WALKER). BM Noct. 8408

D i a g n o s i s : Body black. Wings black, slightly shot with blue. Fore wings with an oblique pale yellow band from just below costa above the upper cell-angle to vein 2 towards the termen, where it ends in a point. Hind wings unspotted. Length of fore wing 20 mm.

D i s t r i b u t i o n : Brazil, Perù, possibly Argentina.

Genus **Neotuerta gen. nov.**

T y p e s p e c i e s : *Agarista sabulosa* FELDER (by desigation hoc locó).

D i a g n o s i s : Proboscis fully developed; antennae slightly dilated towards the extremity; palpi obliquely upturned, the last joint roughly scaled; frons with a truncate conical prominence with a raised terminal rim; sometimes the lower edge of the prominence produced to a point; tibiae fringed with hair above; abdomen with dorsal crests. Venation: in fore wings, vein 2 given off at about three-quarters of cell; 3, 4 and 5 approximated, equidistant; 6 from the lower cell-angle; areole narrow; 7 and the short stalk of 8 + 9 short stalked from the extremity of the areole; 10 from the upper margin of the latter, near the extremity. In hind wings, vein 2 given off beyond three-quarters of cell; 3, 4 and 6, 7 from a point; 5 obsolescent, from middle of the discocellulars.

Male genitalia: Uncus slightly shorter than half of valva, narrow, slightly arched, with a minute terminal hook. Tubus analis broad, well sclerified both dorsally and ventrally. Tegumen moderately broad, with a well developed bunch of stiff hair. Valva banana-shaped; apex with a variable row spinules; fold of sacculus rather broad, nearly or quite as broad as the valva, ending in a slender arched harpe. Aedeagus moderately robust, nearly straight, slightly longer than half of costa, bearing terminally a small patch of minute warts (two such patches in *N. lycaon* [DRUCE]); fultura inferior rather elongate, in most species (but not in *N. sabulosa* [FELDER]) broadening proximally. Saccus short, squarish or rounded.

The American species are generally placed in the Old world genus *Tuerta* WALKER (type species *T. chrysochlora* WALKER), but differ quite distinctly in the male genitalia.

D i s t r i b u t i o n : West Indies, Central and South America.

Key to the species

1 Frontal prominence with its lower edge produced to a point . . . . *N. platensis*, p. 65
1' Frontal prominence without produced lower edge 2
2 Ground colour of the fore wings chocolate grey with some white sprinkling *N. lycaon*, p. 66
2' Ground colour of the fore wings whitish or pale olive 3
3 Ground colour of the fore wings pale olive sprinkled with brownish-red *N. hemicycla*, p. 66
3' Ground colour of the fore wings whitish, thickly sprinkled with chocolate red and silvery . . . . *N. sabulosa*, p. 63

Neotuerta sabulosa (FELDER)

(fig. 21)

1874 *Agarista sabulosa* FELDER, Reise Novara, **107**:11. – T e r r a t y p i c a : San Francisco, California (Type in British Museum).
1874 ? *Eusemia sabulosa* – BOISDUVAL, Rev. Zool., (3) **2**:106.
1875 *Metagarista sabulosa* – BUTLER, Ann. Mag. Nat. Hist., (4) **15**:139.

1882 *Fenaria sevorsa* GROTE, Papilio, **2**:132. – Terra typica: San Francisco, California, (ex errore, fide TODD 1966).
1890 *Agarista noctuiformis* MÖSCHLER, Abhandl. Senckenberg Ges., **16**:112. – Terra typica: Porto Rico (Type in Berlin Museum).
1894 *Seudyra sabulosa* – NEUMOEGEN & DYAR, J. New York entomol. Soc., **2**:19.
1896 *Copidryas sabulosa* – DRUCE, Biol Centr. Amer., Heterocera, **2**:329.
1901 *Tuerta sabulosa* – HAMPSON, Cat. Lepidopt. Phal. Brit. Mus., **3**:625.
1903 *Misa sabulosa* – DYAR, Bull U.S. Nation. Mus., **52** (Nr. 96):96.
1919 *Tuerta sevorsa* forma *albidisca* DRAUDT, in SEITZ, Großschmetterl. Erde, **15**:5. – Terra typica: not designated.
1919 *Tuerta sevorsa* forma *albiplaga* DRAUDT, in SEITZ, Großschmetterl. Erde, **15**:5. – Terra typica: not designated.
1924 *Tuerta noctuiformis* – WOLCOTT, Porto Rico Dept. Agricult. J., **7**:160.
1909 *Tuerta insulica* HAMPSON, Ann. Mag. Nat. Hist., (8) **4**:365. – Terra typica: Trinidad (Type, ♂, in British Museum).

Diagnosis: Head white; palpi white with the sides of the second and the last joint blackish; antennae blackish; thorax clothed with white and dark chocolate brown scales, but sometimes nearly whitish; fore tibiae and the tarsi banded with black; abdomen orange with dark dorsal crests; underside white. Fore wings white, densely sprinkled with chocolate brown and silvery scales, with sometimes almost pure white patches at base and before the costal half of the postmedial line; a dark antemedial line, very oblique from costa to the submedian fold where it is obtusely angled; cell- and discocellular marks blackish with silvery scales on them, the

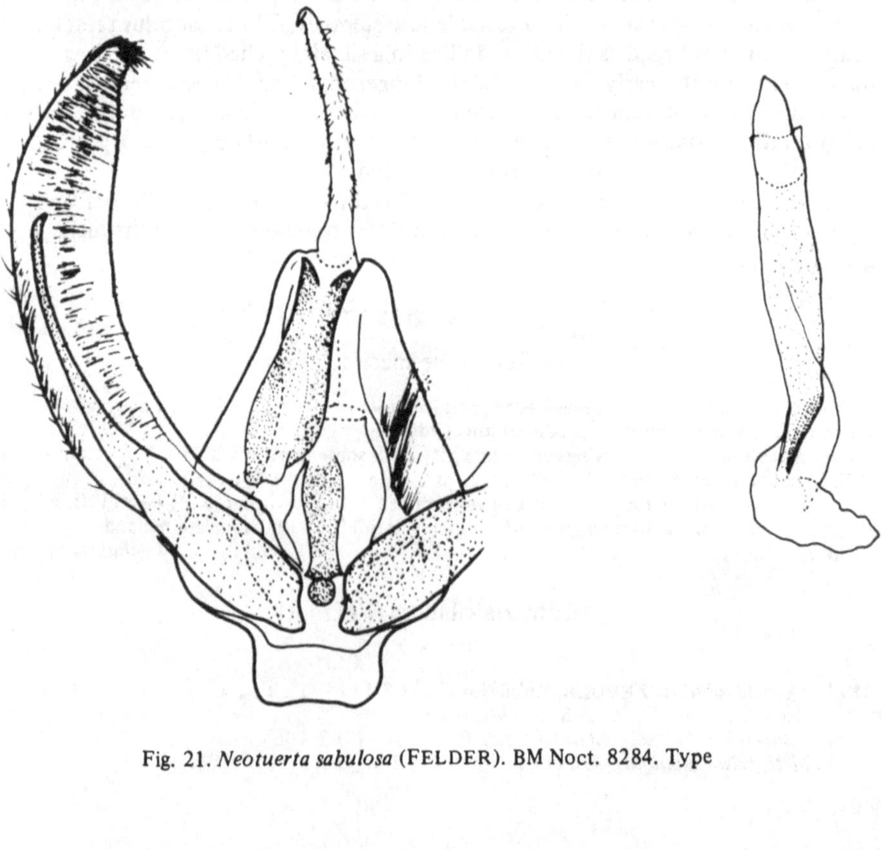

Fig. 21. *Neotuerta sabulosa* (FELDER). BM Noct. 8284. Type

latter produced to points above and below; postmedial line strongly angled inwards at vein 5 and strongly incurved below vein 3; a dentate subterminal line angled inwards at vein 2 and with patches of silvery scales before it; cilia chequered white and blackish. Hind wings orange yellow; terminal black band wide at costa, narrowing to the anal angle; cilia white, blackish above the anal angle. Length of fore wing 17 mm (♂), 19 mm (♀).

Distribution: Mexico, Guatemala, Porto Rico, El Salvador, Trinidad, Cuba.

Neotuerta sabulosa sabulosa (FELDER)

Diagnosis: Vide that of the species.

Distribution: Mexico, Guatemala, Porto Rico, El Salvador, Trinidad.

Neotuerta sabulosa collectiora (TODD)

1966 *Tuerta sabulosa collectiora* TODD, Proc. U.S. Nat. Museum, 120:13, fig. 14, 15. – Terra typica: Baracoa, Cuba (Type, ♂, in U.S. National Museum).

Diagnosis: Proximal margin of the black terminal band on the hind wings excavates in cells M_3 and Cu_1.

Distribution: Cuba.

Neotuerta platensis (BERG)

1882 *Euthisanotia platensis* BERG, An. Soc. Ciencia Argent., **13**:172. – Terra typica: Buenos Aires, Argentina (Type, ♂, in Museum Buenos Aires).
1892 *Copidryas platensis* – KIRBY, Syn. Cat. Lepidopt. Heterocera,:39.
1901 *Tuerta platensis* – HAMPSON, Cat. Lepidopt. Phal. Brit. Mus., **3**:622; fig. 273.

Diagnosis: Head and thorax clothed with chocolate brown and grey hair; tarsi blackish ringed with white; abdomen yellow, mottled with rufous; base and underside whitish; dorsal crests dark. Fore wings whitish, thickly sprinkled with chocolate brown; a white fascia from the base in and below the cell, indented by the cell- and the discocellular spots, then bent up to costa before the apex; area below it chocolate brown tinged with olive; cell-spot and discocellular mark olive edged with black, and with leaden centres; a terminal series of black lunules. Hind wings yellow; a broad terminal reddish-brown band, angled inwards in interspace I; a yellow terminal spot on it above the anal angle; cilia white. Length of fore wing 15 mm.

Distribution: Argentina, Uruguay, Venezuela.

Neotuerta platensis platensis (BERG)

1908 *Tuerta platensis* ssp. *platensis* – JORDAN, Novit. Zool., **15**:254.

Diagnosis: Vide that of the species.

Distribution: Argentina and Uruguay.

Neotuerta platensis bolivar (JORDAN)

1908 *Tuerta platensis* ssp. *bolivar*. JORDAN, Novit. Zool., **15**:254. – Terra typica: Bolivar, Venezuela (Type, ♀, in British Museum).

Diagnosis: Fore wings with the white fascia broader; cell-spot large elongate; discocellular spot larger. Dark markings on the abdomen less distinct.

Distribution: Venezuela.

Neotuerta hemicycla (HAMPSON)

1904 *Tuerta hemicycla* HAMPSON, Ann. Mag. Nat. Hist., (7) **14**:166. – Terra typica: Abaco, Bahamas (Type, ♂, in British Museum).

Diagnosis: ♂: Head and thorax clothed with reddish-brown and white hair; tarsi ringed with white; abdomen fulvous, tinged with red above, and whitish at the base. Fore wings pale olive, thickly sprinkled with reddish-brown; a subbasal red band from costa to the submedian fold; cell- and discocellular spots defined by red, with red centres, and with patches of silvery scales on them; a semicircular white mark defining the cell-spot and the proximal side of the discococellular mark; distal side thereof defined by an oblique white band from below costa to below vein 4 with silver between them from the semicircular white mark to the dorsum; a fine terminal line; cilia spotted red and white. Hind wings pale reddish, with deep red terminal area; cilia spotted red and white. Length of fore wing 13 mm. ♀: Fore wings broader, with the apex less acute; antennae filiform. Fore wings white, sprinkled with dark brown in the terminal area; middle part of costa, terminal area and cell- and discocellular marks tinged with greenish; lines black; area between the ante- and postmedial lines metallic grey in dorsal area; terminal line black, with dark brown spots above the tornus and in interspace V. Hind wings orange-yellow with a dark brown terminal band; discocellular mark small and rather indistinct; terminal line black. Length of fore wing 14–16 mm.

Distribution: The Bahama Islands, Cuba, Florida.

Neotuerta lycaon (DRUCE)

1897 *Euthisanotia lycaon* DRUCE, Ann. Mag. Nat. Hist., (6) **20**:300. – Terra typica: Ecuador, Sarayacu (Type, ♂, in British Museum).
1908 *Tuerta lycaon* – JORDAN, Novit. Zool., **15**:254; pl. 9, fig. 1.
1920 *Paraegocera lycaon* – HAMPSON, Cat. Lepidopt. Phal. Brit. Mus., Suppl., **2**:578; fig. 109.

Diagnosis: Head white; palpi chocolate-brown banded with white; frons brown at the sides; antennae brown; thorax chocolate-red with broad white stripes on the collar and the tegulae, and the middle of thorax; pectus yellowish-white; legs brown, with the tibiae and the tarsi ringed with white; abdomen orange-yellow; basal crest chocolate-red; a dorsal series of diffused brown spots; underside whitish. Fore wings chocolate-brown-grey, sparsely sprinkled with white scales, except at the base; subbasal line white, very oblique from costa to vein 1, where it joins the oblique white antemedial band, arising just below costa and ending at vein 1; cell-spot represented by an oblique silvery bar surrounded, except above, by a large U-shaped white mark; discocellular mark represented by an oblique silvery bar; a large white patch beyond the cell from just below costa to vein 4; a slight, diffused whitish bar beyond it from costa; cilia white at the extremities. Hind wings orange-yellow with a dark chocolate-brown terminal band, narrowing to the anal angle; cilia white at the extremities. Length of fore wing 15 mm.

D i s t r i b u t i o n : Ecuador, Upper Brazil (Amazonas).

Genus **Melanchroiopsis** DYAR

1854 *Phasis* WALKER, List Lepidopt. Ins. Brit. Mus., **2**:312, nec HÜBNER, Verzeichn. exot. Schmetterl. – S p e c i e s t y p i c a : *Josia noctilux* WALKER (by monotypy).
1918 *Melanchroiopsis* DYAR, Proc. U.S. Nation. Mus., **54**:341. – S p e c i e s t y p i c a : *Melanchroiopsis acroleuca* DYAR (by original designation).

D i a g n o s i s : Proboscis fully developed; antennae moderately dilated towards the extremity; palpi with the second segment oblique, sparsely fringed in front with long hair; the last segment moderate, porrect, roughly scaled; frons with a long truncate conical projection, with a small raised rim at the extremity. Nervation: in fore wings, vein 2 arising from about three-quarters of cell; 3, 4 and 5 well separated, equidistant; discocellulars well incurved; 6 from the upper cell-angle; areole small, short; 7 and the stalk of 8 + 9 given off separately from the extremity of the areole; 10 from the upper angle thereof. In hind wings, vein 2 arising from six-sevenths of cell; 3 slightly farther from 2 than from 4; discocellulars incurved; 5 obsolescent, from middle thereof; 6 and 7 from a point.

Male genitalia: Uncus relatively very small, slender, nearly straight, very faintly upturned distally. Tubus analis broad basally, narrowing distally, with ventral sclerification. Vinculum narrow, with a hairy rounded broadening after the middle. Valva relatively much elongate, rather narrow, especially basally, costa and sacculus parallel; apex upturned; fold of sacculus moderately broad, occupying the proximal two-thirds of valva, and ending in a long and slender harpe, with two or three minute subbasal teeth, and a rounded broadening before the hooked terminal part. Aedeagus about one-half of the valva in length, more or less boomerang-shaped; distal portion covered for the most part with minute spicules; fultura inferior narrowing distally; proximal margin broadly rounded; distal edge slightly broader, with the angles prominent. Saccus short, rounded.

D i s t r i b u t i o n : Mexico, Brazil.

Key to the species

1 Head orange . . . . *M. acroleuca*, p. 67
1′ Head black 2
2 Palpi at base, throat and spots on pectus orange . . . . . . *M. noctilux*, p. 68
2′ Head and pectus without orange . . . . *M. mardava*, p. 69

Melanchroiopsis acroleuca DYAR

1918 *Melanchroiopsis acroleuca* DYAR, Proc. U.S. Nation. Mus., **54**:341. – T e r r a t y p i c a : Mexico, Sierra del Guerrero (Type, ♂, in U.S. National Museum).

D i a g n o s i s : Head orange; antennae black; palpi orange, with a black patch at the side of the second segment, and the last segment black; thorax and abdomen black glossed with blue; tegulae orange at the extremities; pectus deep orange; anal tuft orange. Fore wings black; veins and the submedian fold streaked with silvery bluish-grey; cell with a similar coloured streak, forking towards its extremity; a rather small white apical patch; cilia silvery grey. Hind wings black glossed with blue, expecially on the veins; cilia white. Length of fore wing 23 mm.

D i s t r i b u t i o n : Mexico.

Melanchroiopsis noctilux (WALKER)

(fig. 22)

1854 *Josia* (sectio *Phasis*) *noctilux* WALKER, List Lepidopt. Het. Brit. Mus., **2**:312. – T e r r a t y p i c a : Amazonas, Tapajos (Type, ♂, in British Museum).
1868–1874 *Agarista batesii* FELDER, Reise Novara,:pl. 107, fig. 8. – T e r r a t y p i c a : South America (Type not designated).
1874 *Eusemia batesii* – BOISDUVAL, Rev. Mag. Zool., (3) **2**:103.
1892 *Phasis noctilux* – KIRBY, Syn. Cat. Lepidopt. Heterocera,:32.
1892 *Phasis batesii* – KIRBY, Syn. Cat. Lepidopt. Heterodera,:32.

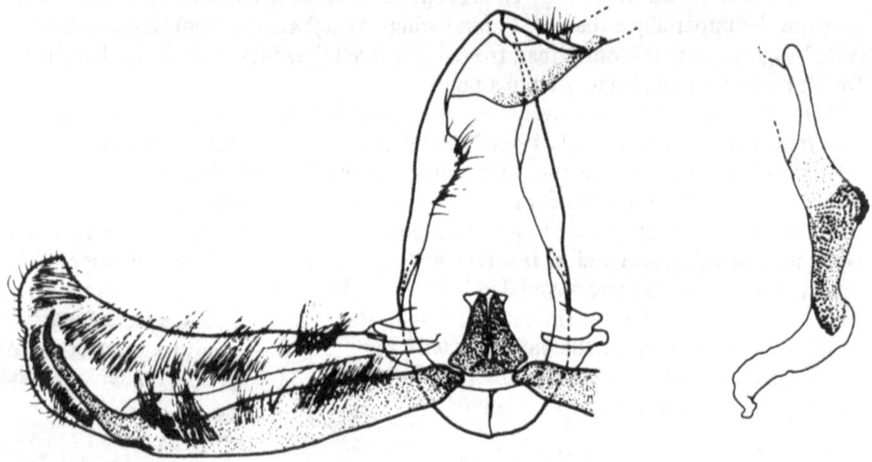

Fig. 22. Melanchroiopsis noctilux (WALKER). BM Noct. 8409. Type

D i a g n o s i s : Body blackish brown; palpi towards the base, postocular streak and lateral spots on pectus in front orange. Wings blackish brown, slightly shot with blue. Fore wings with a pale yellow band from just below costa above the upper cell-angle to the submedian fold near the termen, its proximal edge excised below the costa. Hind wings unspotted. Length of fore wing 22 mm.

D i s t r i b u t i o n : Brazil.

Melanchroiopsis mardava (DRUCE)

1897 *Phasis mardava* DRUCE, Ann. Mag. Nat. Hist., (6) **20**:300. – Terra typica: Amazonas, Rio Napo (Type, ♀, in British Museum).

Diagnosis: Body and wings blackish brown. Fore wings with a slightly irregular pale yellow band from just below middle of costa to below vein 2 above tornus, incurved below the costa. Length of fore wing 22 mm.

Diagnosis: Brazil.

Genus Caularis WALKER

1857 *Caularis* WALKER, List. Lepidopt. Het. Brit. Mus., **12**:800.– Species typica: *Caularis undulans* WALKER (by monotypy).

Diagnosis: Proboscis fully developed; antennae of male bipectinate with rather long branches to apex; palpi moderately long, obliquely porrect; second joint fringed with long hair below; third joint moderate; frons with a large truncate conical prominence, with a small process at the extremity with the rim raised; eyes large, rounded; tibiae fringed with long hair; thorax clothed with long hair, but without crests; abdomen with a large dorsal crest on the first urite. Nervation: in fore wings, vein 2 from seven-tenths of cell; 3 and 4 much approximated; 4 and 5 from nearly a point; discocellulars well incurved; 6 from the upper cell-angle; areole narrow; 7 and the long stalk of 8 + 9 from the extremity of areole; 10 from upper margin of areole. Hind wings with the termen slightly produced and angled between veins 2 and 4; vein 2 from slightly beyond two-thirds of cell; 3, 4 and 6, 7 from the cell-angles; discocellulars faintly angled at middle; 5 obsolescent, from middle of discocellulars.

Male genitalia: Uncus slightly longer than one-half of costa, compressed, arched before one-third, then nearly straight, ending in a minute hook. Tubus analis rather slender, sclerified only ventrally. Vincula broadly fused distally; fold moderate, bearing moderately long, stiff pilosity. Valva very elongate, more or less banana-shaped but with the sacculus distinctly angled at three-fifths; fold rather narrow, reaching to three-fifths of valva, ending in a short, sickle-shaped harpe. Aedeagus relatively large and robust, as long as costa, on the whole boomerang-shaped, with the proximal half spoon-shaped; a rather large subterminal and terminal ventral patch of small cornuti; vesica very elongate and narrow, unarmed; fultura inferior with the proximal margin narrowly excised at middle, and with the angles produced; distal part rather elongate and narrow; terminal margin cut straight; fultura superior very broad, with the sides narrowly sclerified. Saccus semi-elliptical, slightly pointed.

Distribution: West Indies, Bahama Islands.

Key to the species

1 Hind wings yellowish white with a large brown discocellular lunule . . . . . *C. lunata*, p. 71
1′ Hind wings orange yellow, without discocellular lunule 2
2 Fore wings with the tornal area bearing a short white streak . . . . *C. undulans*, p. 70
2′ Fore wings with the tornal area bearing a white V-mark . . . . *C. jamaicensis*, p. 71

Caularis undulans WALKER

(fig. 23)

1857 *Caularis undulans* WALKER, List Lepidopt. Het. Brit. Mus., **12**: 801. – Terra typica: Haiti (Types, ♂, ♀, in British Museum).
1874 *Eudryas bartholomaei* BOISDUVAL, Rev. Mag. Zool., (3) **2**:59. – Terra typica: St. Barthelemy Is.

Diagnosis: Head white mixed with fuscous; second palpal joint black except in front and at extremity; sides of frons marked with black; antennae rufous; thorax olive-brown mixed with greyish; pectus and legs yellowish white, latter sprinkled with fuscous; fore tarsi blackish ringed with white; abdomen pale yellow; basal crest large, black mixed with metallic blue scales laterally, but white medially. Fore wing whitish to the postmedian line, sprinkled with olive-brown, especially costally and dorsally; postmedian area olive-brown; terminal area greyish, thickly sprinkled with olive-brown; a large antemedial olive-brown patch on dorsum; cell- and discocellular marks small, ill-defined, olive-brown with some silvery-blue in centres, the former rounded; postmedial line dark, from costa at three-quarters, strongly incurved below vein 5; a white fascia in the submedian fold from the postmedian area to termen; a white terminal line; cilia blackish tipped with silvery-blue

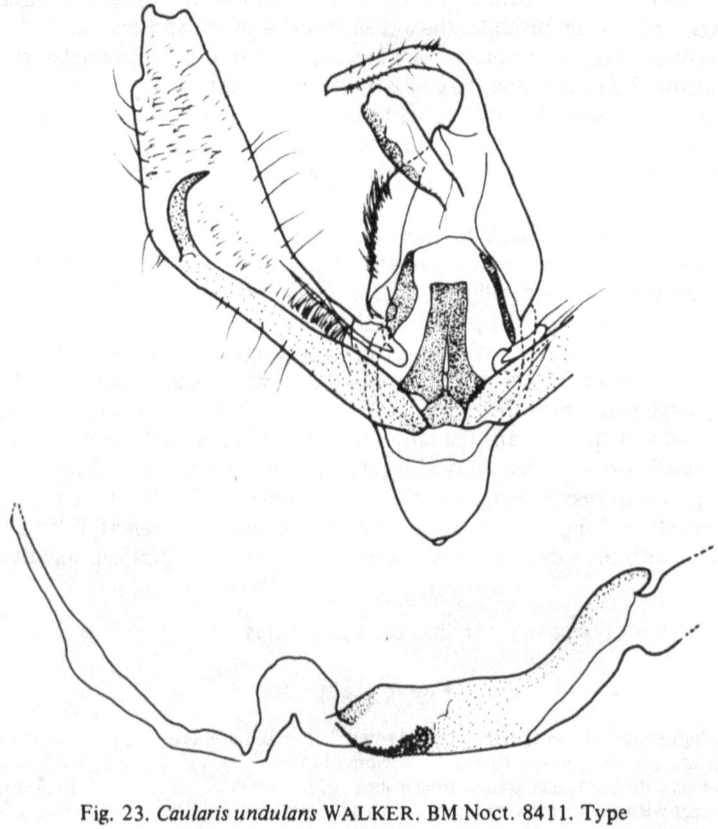

Fig. 23. *Caularis undulans* WALKER. BM Noct. 8411. Type

from tornus to vein 2. Hind wings pale yellow with a salmon terminal line; a black spot on it below vein 3 and a streak below vein 2, ending at vein 1; a fine dark terminal line; cilia whitish, but blackish from vein 1 to vein 3. Length of fore wing 14–22 mm.

Distribution: Santo Domingo, Jamaica, Puerto Rico, Virgin Island.

Caularis lunata HAMPSON

1904 *Caularis lunata* HAMPSON, Ann. Mag. Nat. Hist., (7) **14**:169. – Terra typica: Bahama Islands, Nassau (Type, ♀, in British Museum).

Diagnosis: Head and thorax red brown mixed with white; sides of frons and pectus white; legs white mixed with brown; fore tarsi banded with brown; abdomen yellowish white with a dorsal series of brown spots; basal crest small, black; three medial urites with lateral black spots; underside tinged with brown. Fore wings white, almost wholly tinged and sprinkled with olive-brown; a broad white fascia from base of costa and thence in the submedian interspace to the postmedial line; antemedial line represented by a diffuse olive bar from the submedial fascia to dorsum; cell- and discocellular marks as obscure olive patches with some silvery scales in centres; extremities of the median vein and the discocellulars with some indian red on them; postmedial line indian red with a white band proximally on costa, towards which it is diffused to the submedian fascia; postmedian area olive; a white fascia in submedian fold from the postmedian area to termen; a fine dark terminal line with a series of white lunules before it. Hind wings yellowish white; a large fuscous discocellular lunule; a pale red terminal band from vein 1 to apex, shading to brown on its proximal edge, with an oblique black spot on it at the submedian fold; a small black spot at vein 1; a fine indian red terminal line; cilia white. Length of fore wing 16–18 mm.

Distribution: Bahama Islands, Cuba.

Caularis jamaicensis TODD

1910 *Caularis undulans* HAMPSON, Cat. Lepidopt. Phal. Brit. Mus., **9**:406; fig. 191, nec. WALKER 1857. – Terra typica: Jamaica.
1966 *Caularis jamaicensis* TODD, Proc. U.S. Nation. Mus., **120**:2–4; fig. 17. – Terra typica: Jamaica (Type, ♂, in U.S. National Museum).

Diagnosis: Head grey; palpi black, mixed with some white; collar grey; tegulae and thorax red-brown, darker behind; abdomen pale yellow, the basal urite with a tuft of blue-black scales; pectus pale yellow or whitish; legs mixed dark and light, the fore legs darker. Fore wings with a basal spot along dorsum, the area around the discocellular and the subcellular marks, and the subterminal area distally of the postmedial line, dull olive-green; postmedial line dark reddish brown; anterior half of the subterminal area suffused with darker reddish brown scales; medial area, including the so-called "typical" spots, and the terminal area suffused with dark grey metallic scales, with some violet iridescence; a narrow V formed by two vague lines above tornus; cilia brown, darker tornally. Hind wings yellow-orange; a vague salmon coloured marginal band from apex to vein 3; interspace III with a black spot; interspace II with a large salmon spot, bordered with black; cilia yellowish or whitish from apex to middle fo interspace III, then metallic black to the anal angle. Length of fore wing 18–19 mm (♂), 19–21 mm (♀).

Distribution: Jamaica.

Genus **Gerra** WALKER

1864 *Gerra* WALKER, List Lepidopt. Het. Brit. Mus., **31**:141. – Species typica: *Gerra radicalis* WALKER (by original designation).
1882 *Fenaria* GROTE, Papilio, **2**:132. – Species typica: *Fenaria sevorsa* GROTE (by monotypy).

Diagnosis: Proboscis fully developed; antennae of male bipectinate, with moderately long branches to apex; antennae of female serrate; palpi obliquely porrect to just beyond frons; the second joint with long hair below, the third joint short. Eyes large, rounded. Frons with a large rounded prominence, with a truncate conical median process, with raised edges, and with a corneous plate below it. Thorax clothed with hair and hair-like scales, without crests; tibiae fringed with long hair; abdomen with dorsal crests on the four or five first urites.

Male genitalia: Uncus about half of costa in length, slender, bent at one third, then more or less straight. Fold of vinculum narrow, roundedly produced at middle, with the pilosity scarce and reduced to the distal portion. Valva banana-shaped; apex with a few rather stiff hairs; a couple of similar hairs on the adjoining portion of the sacculus; base bearing subcostally a short process covered terminally with minute spinules; fold of sacculus rather narrow, hardly half as broad as the valva at the base, reaching to the middle of the latter, and ending in a short, slender, arched harpe. Aedeagus about three-fourths of costa in length; the proximal two-thirds spoon-shaped; the distal third very robust; terminal margin with 3 or 4 rows of minute cornuti; the everted vesica as long as the body of the aedeagus; fultura inferior rather weakly developed, but possibly damaged. Saccus semi-elliptical, more or less pointed.

Distribution: U.S.A., Central and South America.

Key to the species

1 Shoulders with an orange patch . . . . *G. radicalis*, p. 73
1′ Shoulders unspotted 2
2 Hind wings orange or yellow with a dark marginal band 3
2′ Hind wings greyish-black . . . . *G. sophocles*, p. 75
3 Hind wings yellow; terminal band blackish-maroon, giving off a streak on the cubital vein . . . . *G. lunata*, p. 75
3′ Hind wings orange; terminal band without streak on cubital vein 4
4 Thorax rufous . . . . *G. aelia*, p. 74
4′ Thorax brown, usually mixed with white hairs or scales 5
5 Fore wings rather elongate, with the termen oblique; ground colour purplish-grey; costal area deep red-brown; numerous oblique, dentate transverse bars . . . . . . *G. pulchra*, p. 75
5′ Fore wings less elongate, with the termen less oblique; ground colour brown, without numerous oblique dentate transverse lines 6
6 Fore wings chocolate-brown mixed with whitish scales; discocellular mark perpendicular to costa . . . . *G. sevorsa*, p. 74
6′ Fore wings greyish-brown, mixed with bronze-coloured scales; discocellular mark perpendicular to dorsum . . . . *G. brephos*, p. 74

Gerra radicalis WALKER
(fig. 24)

1884 *Gerra radicalis* WALKER, List Lepidopt. Het. Brit. Mus., **31**:141. – Terra typica: ignota (Type, ♂, in British Museum).
1889 *Diamuna adrasta* DRUCE, Biol Centr. Amer., Heterocera, 1:334; pl. 30, fig. 20. – Terra typica: Patycuaro (Type, ♀, in British Museum).

Diagnosis: Head and thorax chocolate-brown with a few metallic-blue scales; shoulders with orange patches; abdomen blackish-brown with some metallic-blue scales on the dorsal crests. Fore wings chocolate-brown with bluish-white and

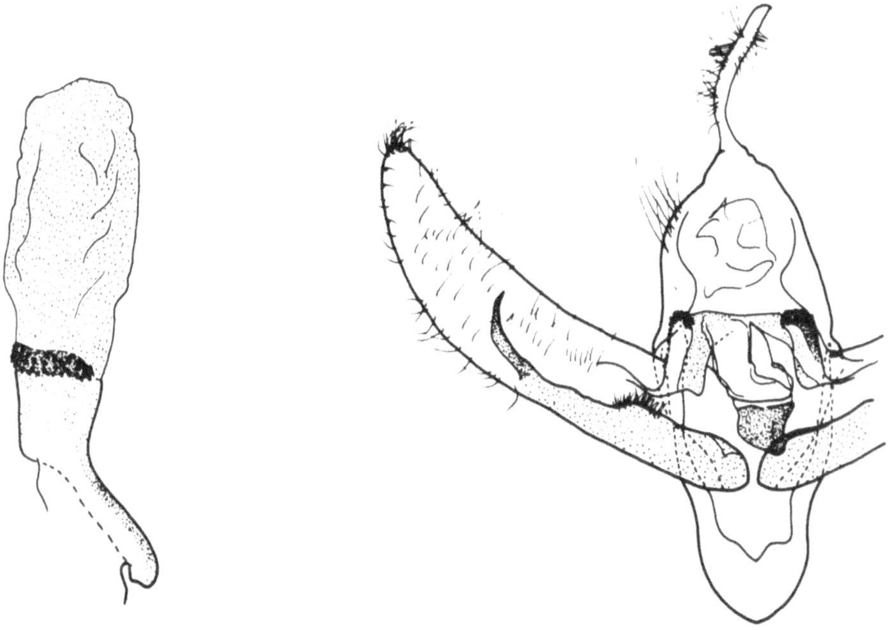

Fig. 24. *Gerra radicalis* WALKER. BM Noct. 4813. Type

metallic silvery-blue scales; a very indistinct rufous antemedial line, oblique to the submedian fold, then angled twice; cell- and discocellular marks edged with rufous and with some metallic-blue scales in centres, the former round, the latter rather irregular in shape, and with a faint yellowish patch beyond it; postmedial line indistinct, minutely waved; subterminal line irregularly sinuous, dark, edged distally by a bright rufous suffusion, then by a slight pale band; a faint pale line at the base of the cilia. Hind wings orange-yellow with a dark brown terminal band; cilia greyish at the extremities. Length of fore wing 19–22 mm.

Distribution: Mexico.

Gerra sevorsa (GROTE)

1882 *Fenaria sevorsa* GROTE, Papilio, **2**:132. – Terra typica: U.S.A., Arizona (Type in U.S. National Museum).
1882 *Phaegorista sevorsa* – GROTE, Papilio, **2**:189.
1883 *Gerra sevorsa* – GROTE, Canad. Entomologist, **15**:5.
1889 *Diamuna aedessa* DRUCE, Biol. Centr. Amer., Heterocera, **1**:334; pl. 30, fig. 21 and 22. – Terra typica: Guatemala, Volcan de Atitlan (Type, ♂, in British Museum).
1919 *Gerra sevorsa* forma *meridionalis* DRAUDT, in SEITZ, Großschmetterl. Erde, **11**:11. – Terra typica: Mexico, Guatemala.
1921 *Gerra sevorsa* aberr. *luteomacula* STRAND, Archiv f. Naturgesch., **11**:165. – Terra typica: not designated.

Diagnosis: Head and thorax chocolate-brown mixed with whitish scales; tarsi blackish ringed with white; abdomen blackish-brown; dorsal crests with some metallic scales; anal tuft with some grey. Fore wings chocolate-brown sprinkled with bluish-white and some metallic silvery scales; a diffused whitish patch beyond the cell from the costa to vein 3; antemedial line indistinct, oblique, waved; cell- and discocellular marks olive-brown with some metallic scales in centres, the former round, edged with chocolate-red, the latter edged with red proximally only; postmedial line very indistinct, diffused; traces of a pale waved subterminal line with rufous marks before and beyond it in the interspaces; cilia with a slight pale basal line. Hind wings orange with a rather broad black terminal band expanding at apex and tornus. Length of fore wing 16–22 mm.

Distribution: U.S.A. (Arizona), Mexico, Guatemala.

Gerra brephos DRAUDT

1919 *Gerra brephos* DRAUDT, in SEITZ, Großschmetterl. Erde, **11**:11; pl. 1h.– Terra typica: Mexico, Zacualpan (Type in Berlin Museum).

Diagnosis: Very similar to *Gerra sevorsa* (GROTE); fore wings greyer, sprinkled with bronze-coloured scales; discocellular mark perpendicular to the dorsum, not to the costa as in *G. sevorsa*; outer transverse line more distinct, dentate, parallel to the termen, not bending around the discocellular mark.

Distribution: Mexico.

Gerra aelia (DRUCE)

1889 *Diamuna aelia* DRUCE, Biol. Centr. Amer., Heterocera, **1**:334; pl. 30, Fig. 23 and 24. – Terra typica: Panama, Chiriqui (Type, ♂, in British Museum).
1910 *Gerra aelia* – HAMPSON, Cat. Lepidopt. Phal Brit. Mus., **9**:418.

Diagnosis: Head, collar, pectus and legs chocolate-brown mixed with some white hairs; thorax rufous; abdomen dark brown, the crests with purplish scales mixed; underside with some grey hairs. Fore wings chocolate-brown, sprinkled with bluish-white scales, especially on the costal area, on veins and beyond the cell; antemedial line indistinct, double, filled in with olive-brown, oblique, waved; cell- and discocellular marks olive-brown edged with chocolate-red, the former rounded. the latter with a chocolate lunule on the discocellular vein; postmedial line chocolate, edged with olive-brown on the distal side; an indistinct diffused curved subterminal chocolate line, and a somewhat lunulate terminal line; cilia olive-brown, chequed with whitish. Hind wings orange; a broad blackish-brown terminal band, expanding on the costa and the anal margin; cilia olive-brown with a slight pale basal line. Length of fore wing 17 mm.

D i s t r i b u t i o n : Panamà, Costa Rica.

Gerra pulchra DRAUDT

1919 *Gerra pulchra* DRAUDT, in SEITZ, Großschmetterl. Erde, **11**:11; pl. 1h. – T e r r a t y p i c a : Bolivia, Mapiri (Type, ♂, in Berlin Museum).

D i a g n o s i s : Fore wings somewhat more elongate with the termen oblique, purplish-grey, sprinkled with silvery-grey scales; costal area deep red-brown; numerous dentate, very oblique transverse lines; postmedial line twice elbowed, and excurved to enclose the discocellular mark; the latter red-brown with the centre silvery-grey, followed by a faint yellowish patch. Hind wings with a blackish-brown terminal band, expanding apically. Length of fore wing 15 mm.

D i s t r i b u t i o n : Bolivia, Colombia.

Gerra lunata KÖHLER

1936 *Gerra lunata* KÖHLER, Physis, **12**:71–72. – T e r r a t y p i c a : Argentina, Jujuy, Mendieta (Type in Nat. Museum Argentina).

D i a g n o s i s : Body light grey, nearly ash-coloured; antennae black with the two basal joints grey; palpi and legs grey; anal tuft light grey. Fore wings grey; veins distinctly paler on a darker ground; a basal whitish spot from the radial and the cubital veins to the anal vein, expanding towards the middle of the wing, with one margin on the median vein, the other between the cubital and the anal veins; a second lunar whitish spot, beginning on the costa before the apex, dilated to near the cubital vein, and ending near the termen above the cubital vein; above the discocellulars a yellowish white point; the whitish markings finely and irregularly edged with yellowish. Hind wings yellow; the termen edged with blackish-maroon, giving off a streak on the cubital vein. Length of fore wing 15 mm.

D i s t r i b u t i o n : Argentina.

Gerra sophocles DYAR

1912 *Gerra sophocles* DYAR, Proc. U.S. Nation. Museum, **1912**:72. – T e r r a t y p i c a : Mexico, Cuernavaca (Type in U.S. National Museum).

D i a g n o s i s : Body brown; proboscis and base of palpi bright orange yellow; abdomen dull greyish black. Fore wings brown with a white oblique costal shade before apex; cell- and discocellular marks lighter brown; inner and outer lines very faint, dark; a light reddish submarginal row of joined spots margined proximally with red brown. Hind wings dull greyish black; cilia white. Length of fore wing 15 mm.

D i s t r i b u t i o n : Mexico.

Genus **Parathisanotia gen. nov.**

T y p e s p e c i e s : *Bombyx grata* FABRICIUS (by designation hoc loco).

D i a g n o s i s : Proboscis fully developed; antennae simple in both sexes (in *Euthisanotia* ciliate in ♂); palpi upturned, with the second segment longer than the last, and densely covered with long scales. Fore femora and tibiae with very dense covering of long hairs (fore tibiae only in *Euthisanotia*). Abdomen with a large dorsal crest on urite I and a smaller tuft on urite II. Fore wings with the termen evenly rounded; hind wings more elongate than in *Epithisanotia* gen. nov., with the costa slightly outcurved, and the termen strongly oblique.

Venation: in fore wings, vein 2 given off from three-quarters of cell; 3, 4 and 5 well separated, equidistant; discocellulars angled inwards; areole short and narrow, somewhat shorter and broader than in *Epithisanotia* gen. nov.; 6 from the upper cell-angle; 7 short stalked with 8 + 9, stalk of 8 + 9 about one-fifth of 8; 10 from upper margin of areole, at about four-fifths of its length. In hind wings, vein 2 arising from three-quarters of cell; 3 and 4 very much approximated; discocellulars strongly incurved; 5 weak, from middle of discocellulars; 6 and 7 from a point; 8 anastomosing with cell at extreme base.

Male genitalia: Uncus short, depressed, with a minute terminal hook. Tubus analis short, broad, well sclerified. Vinculum broad, fused distally, without hair-fringe. Valva elongate, moderately broad, broadening and slightly downcurved apically; costa forming a rounded angle before the middle; fold of sacculus very broad, nearly reaching costa in the basal portion, narrowing distally, reaching beyond two-thirds of valva, and ending in a harpe of a very unusual shape, broad, with lateral processus, and asymmetrical; just in middle of valva, a transversal short, sickle-shaped processus. Aedeagus of the type found in *Euthisanotia*, with a very short broad base and two long and slender, nearly filiform processus, the longer of which about three-quarters of costa in length; fultura inferior narrow, much produced distally. Saccus very broad, rounded.

D i s t r i b u t i o n : North America.

Parathisanotia grata (FABRICIUS)

(fig. 25)

1793 *Bombyx grata* FABRICIUS, Syst. entomol., 3 (1);457. – T e r r a t y p i c a : North America (Type not designated).
1872 *Eudryas grata* – STRETCH, Zygaenae and Bombyces North America,:147; pl. 7, fig. 1.
1874 *Eudryas assimilis* BOISDUVAL, Rev. Mag. Zool., (3) 2:59. – T e r r a t y p i c a : U.S.A., Georgia (Type in British Museum).
1892 *Euthisanotia grata* – KIRBY, Syn. Cat. Lepidopt. Heterocera,:39.
1893 *Euthisanotia cypris* GROTE, Canad. Entomologist, 25:322. – T e r r a t y p i c a : Assumption.

D i a g n o s i s : Head and thorax above blackish-brown with some silvery-blue scales; tegulae white; pectus and base of legs white; rest of legs blackish-brown;

fore tibiae fringed with brown mixed with white; tarsi ringed with white. Abdomen orange, with dorsal and lateral series of black spots; anal tuft black preceded by some white; underside white. Fore wings white; costal area indian red to the cell-extremity, and extending to the discal fold, sprinkled with white and becoming

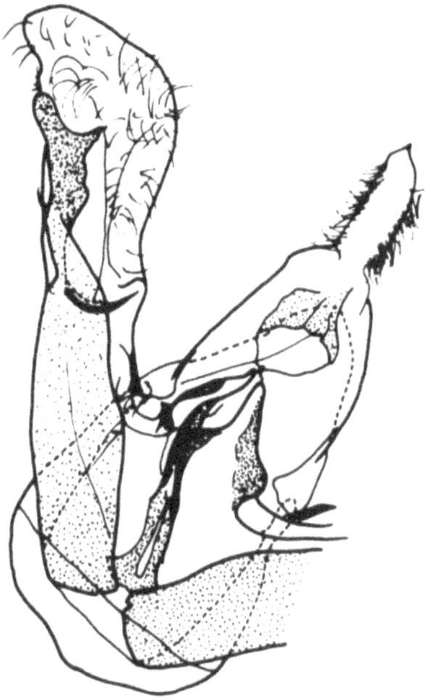

Fig. 25. *Parathisanotia grata* (FABRICIUS). BM Noct. 8487

blackish at costa; cell-spot deep indian red, rounded; discocellular spot indian red and blackish above, its lower part olive, much reduced; a brownish-olive, speckled with silvery-blue patch on middle of dorsum, extending to near base; postmedial line very distal, olive with the proximal edge darker and browner; area beyond it indian red, shading to yellowish on termen; a waved whitish subterminal line with brown mark on it at middle, and two blackish marks beyond it above termen; a terminal series of indian red lunules edged with yellow. Hind wings orange-yellow; a faint dark discocellular spot; terminal band indian red with waved inner edge, with a yellow striga on it towards tornus and a black striga above it on anal area; cilia tipped with white. Length of fore wing 13–18 mm.

E a r l y s t a g e s : Larva: Head pale reddish with six small dots of black on each lobe. Body whitish, each segment with a median transverse pinkish stripe containing black dots or tubercles; rest with six transverse purplish lines; shields and legs pale pinkish with black dots. On Ampelopsis.

D i s t r i b u t i o n : Most parts of North America.

Genus **Euthisanotia** HÜBNER

1827–1831 *Euthisanotia* HÜBNER, Zutr. exot. Schmetterl., **3**:12, 39. – S p e c i e s t y p i c a : *Euthisanotia unio* HÜBNER (by subsequent designation: KIRBY 1892).

1836 *Eudryas* BOISDUVAL, Spéc. gén. Lépidopt., **1**:4. – S p e c i e s t y p i c a : *Euthisanotia unio* HÜBNER.

D i a g n o s i s : Proboscis fully developed; antennae ciliated; palpi upturned, the second joint reaching about the middle of frons, moderately fringed with scales in front; the last joint short; frons with a large rounded prominence, with raised rim at the extremity, and a corneous plate below it; thorax clothed with rough hair and scales mixed; metathorax with a spreading crest; fore tibiae with very large tufts of hair on the outer side; mid and hind tibiae moderately fringed with hair; abdomen with dorsal crests on the basal urites.

Neuration: in fore wings, vein 2 arising from seven-ninths of cell; 3 to 5 approximated, equidistant; discocellulars incurved; 6 from upper cell-angle; areole narrow; 7 and the short stalk of 8 + 9 from its extremity; 10 from its upper angle. In hind wings, vein 2 arising from about middle of cell; 3, 4 and 6, 7 from cell-angles; discocellulars faintly incurved; 5 obsolescent, from middle thereof; 8 anastomosing with cell for a while near base.

Male genitalia: Uncus slightly shorter than one-half of valva, narrow, depressed, nearly straight, with a terminal hook. Tubus analis broad, sclerified, with the base bifid. Vinculum moderately broad, without stiff hair. Valva rather broad, with the margins parallel; costa straight for two-thirds, then suddenly narrowing to a point; sacculus bearing at middle a short process, with a few stiff hairs near its base; narrower apical part of the valva bearing numerous stiff hairs directed costad; fold of sacculus broad basally, but soon narrowing to form a rather robust, faintly arched harpe, which reaches to four-fifths of the valva. Aedeagus consisting almost wholly of two long and very slender, nearly filiform processes, variously bent, and ending in a small club; basal portion very short, rather narrow; total length more than twice that of the valva; fultura inferior narrowing distally, deeply cleft at the middle, with the sides faintly serrated in the distal third. Saccus broadly semicircular.

D i s t r i b u t i o n : North America.

Key to the species

Hind wings pale yellow, without black discal lunule . . . . *E. unio*, p. 78
Hind wings orange-yellow, with a black discal lunule . . . . *E. brevipennis*, p. 80

Euthisanotia unio HÜBNER

(fig. 26)

1827–1831 *Euthisanotia unio* HÜBNER, Zutr. Sammlg. exot. Schmetterl., **3**:12; pl. 431, fig. 2. – T e r r a t y p i c a : U.S.A. (Type not designated).
1836 *Eudryas unio* – BOISDUVAL, Spéc. gén. Lépidopt., **1**:4.

D i a g n o s i s : Head and a broad dorsal fascia on the thorax dark brown, the latter with a few silvery scales mixed; outer part of the collar and tegulae pure white; pectus white; legs brown and white; fore tibiae black mixed and fringed with white; tarsi with slight white rings; abdomen yellow with dorsal and lateral series of blackish spots except towards the extremity; basal crest large, dark brown, mixed with silvery scales; anal tuft and underside white. Fore wings pure white; costal area rusty-red to the cell-end, extending to the discal fold, sprinkled with bluish-white, and with some yellowish on its lower edge; a triangular rusty-red patch

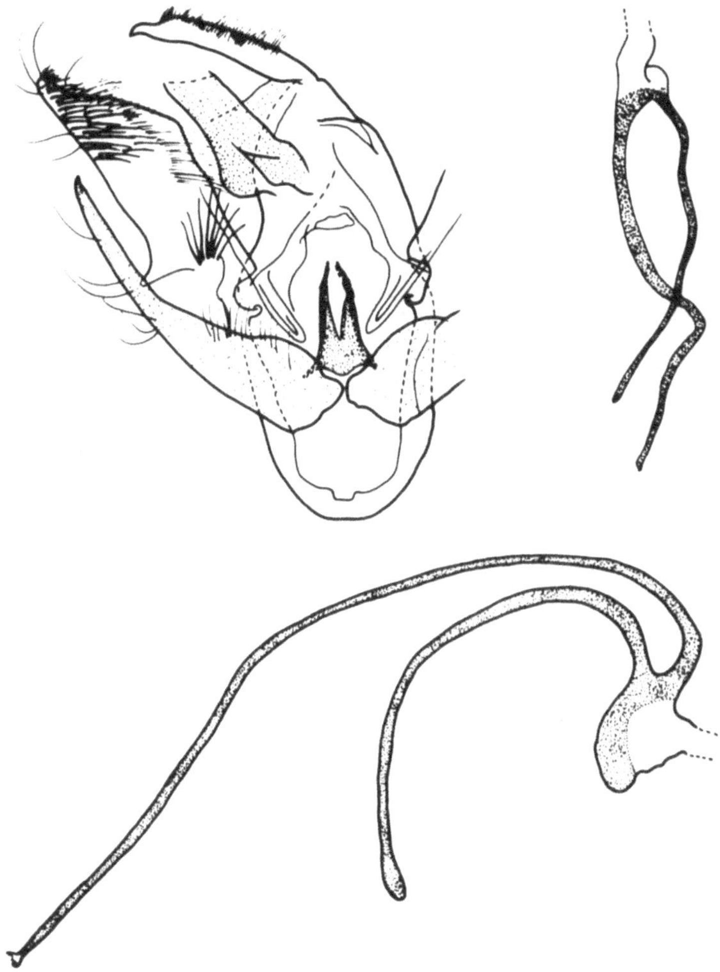

Fig. 26. *Euthisanotia unio* HÜBNER. BM Noct. 8492

on the dorsum from near the base to near the postmedial line, sprinkled with white, and with some diffused olive-yellow on its edges; cell-spot semicircular, olive with slight white annulus; discocellular spot with its upperpart olive with white inner ring, its lower part yellowish with a slight olive lunule on its lower edge; postmedial line fine, white, edged on each side with olive, somewhat oblique to vein 6, then obliquely excurved to vein 2, then slightly incurved; area beyond it rusty-red with an indistinct curved and waved whitish subterminal line; terminal area browner, sprinkled with bluish-white. Hind wings pale yellow, with a rusty-red terminal band from vein 6 to near the anal angle, with slight line before its proximal edge which expands widely in the submedian interspace; a bluish-white lunule near the termen towards the anal angle; cilia white towards the apex, reddish-brown tipped with white towards the anal angle. Length of fore wing 17–22 mm.

E a r l y s t a g e s : Larva: Head red brown with six black spots on each lobe. Body slightly enlarged at the twelfth segment; each segment with a median transverse orange stripe containing black dots, the rest with fine transversal purplish lines on a pale yellow ground; shields and legs orange, dotted with black. On Epilobium.

D i s t r i b u t i o n : Eastern part of the U.S.A.

Euthisanotia brevipennis (STRETCH)

1874 *Eudryas brevipennis* STRETCH, Zygaenae and Bombyces of North America,:151; pl. 7, fig. 3 and 4. – T e r r a t y p i c a : California, Los Angeles (Type, ♀, in U.S. National Museum).
1892 *Euthisanotia brevipennis* – KIRBY, Syn. Cat. Lepidopt. Heterocera,:39.

D i a g n o s i s : Head and thorax very dark chocolate-brown with a few silvery-blue scales; antennae ringed with white; tegulae and pectus white; legs dark brown, with bases and bands white; fore tibiae fringed with white; abdomen orange-yellow with black dorsal crests; basal crest with some of hairs tipped with silvery-blue; underside white with sublateral series of dark brown bars; extremity and anal tuft dark brown. Fore wings purplish-pink; costal edge fuscous to the postmedial band; costal area sprinkled with silvery-blue; inner medial area blackish brown with a white patch on the dorsum; a broad pure white fascia from the base where it extends to the costa and the dorsum, in and below the cell to its lower angle, then obliquely bent upwards to the costa at the postmedial band; a subbasal black dot on it below the cell; costal margin of the fascia indented by the cell- and discocellular mark olive-brown sprinkled with black; lower edge of the fascia edged with olive-yellow to the middle where there is a triangular indentation; its distal edge limited distally by the double, waved blackish postmedial line, filled in with olive-brown, and with a slight white bar on it on its dorsal (i.e. lower) margin; subterminal line curved, waved, whitish, angled inwards to the postmedial line at the discal fold, with slight dark marks beyond it below the costa, at the discal fold and below veins 2 and 3; a terminal series of indistinct lunules edged sith slight white sprinkling proximally; cilia pale at the base, with rufous extremities. Hind wings orange-yellow; a black discocellular lunule; terminal band purplish-red with waved proximal edge; a blackish and white bar at the anal margin, and a slight whitish bar before the termen in the submedian fold. Length of fore wing 17 mm.

E a r l y s t a g e s : Larva: Head red brown with six black spots on each side; body white, slightly enlarged at the 13th segment; each segment with a medial transverse dark orange stripe containing black dots; rest with fine transversal black lines; cervical shield nearly whitish, others orange with black dots.

D i s t r i b u t i o n : U.S.A. (California).

Genera incertae sedis

Genus Pseudalypia HY. EDWARDS

1874 *Pseudalypia* HY. EDWARDS, Proc. Calif. Acad. Sci., **5**:101. – S p e c i e s t y p i c a : *Pseudalypia crotchii* HY. EDWARDS (by original designation).

D i a g n o s i s : Proboscis fully developed; head small, clothed with long hair; antennae hardly dilated towards the extremity; palpi short, porrect, fringed with long hair; frons with a long triangular corneous process notched at sides; eyes small. Thorax hairy; tibiae smoothly scaled. Nervation: in fore wings, vein 3 arising from near the angle of cell; 5 from above the angle; 6 below the upper cell-angle; 9 from 10 anastomosing with 8 to form an areole; 7 from the extremity thereof; 11 from cell. In hind wings, vein 2 arising from well before the lower cell-angle; 3 and 4 from angle; 5 obsolescent, from middle of the discocellulars; 6 and 7 from the upper cell-angle.

D i s t r i b u t i o n : Southern U.S.A.

Pseudalypia crotchii HY. EDWARDS

1874 *Pseudalypia crotchii* HY. EDWARDS, Proc. Calif. Acad. Sci., **5**:111. – T e r r a t y p i c a : California, San Diego (Type, ♀, in American Museum Nat. Hist.).
1884 *Pseudalypia crotchii* var. *atrata* HY. EDWARDS, Papilio, **4**:121. – T e r r a t y p i c a : California.

D i a g n o s i s : Body and wings brownish-black; base of palpi, collar and anal tuft orange-yellow. Fore wing with costa creamy-white from base to the narrow postmedial band, which tapers to a point above dorsum, and is sometimes absent (var. *atrata*, ♂); cilia white at the apex. Hind wings unspotted; cilia white. Length of fore wing 11 mm.

D i s t r i b u t i o n : U.S.A. (Southeastern California).

Genus Letaba DYAR

1912 *Letaba* DYAR, Proc. U.S. Nation. Mus., **42**71–72. – S p e c i e s t y p i c a : *Letaba noa* DYAR (by original designation).

D i a g n o s i s : Palpi oblique, densely scaled, the terminal segment short; frons with a truncate conical prominence, flattened at the extremity, with the edges scarcely raised; eyes large, round. Thorax and abdomen without crests; fore legs without spines. Fore wings with an areole. Nervation? (not seen).

D i s t r i b u t i o n : Mexico.

Letaba noa DYAR

1912 *Letaba noa* DYAR, Proc. U.S. Nation. Mus., **42**:72. – T e r r a t y p i c a : Mexico, Zacualpan (Type, ♀, in U.S. National Museum).

D i a g n o s i s : Body black; proboscis and basal segment of palpi bright orange yellow. Wings black; cilia of both wings white. Underside of fore wings with base and costal line, and underside of hind wings with basal half of costa bright orange yellow; cilia white. Length of fore wing 19 mm.

D i s t r i b u t i o n : Mexico.

Genus **Eupseudomorpha** DYAR

1880 *Edwardsia* NEUMOEGEN, Canad. Entomologist, **12**:67; nec QUATREFAGES, Pol. 1842. – Species typica: *Edwardsia brillians* NEUMOEGEN (by monotypy).
1893 *Eupseudomorpha* DYAR, Canad. Entomologist, **25**:28.

Diagnosis: Proboscis fully developed; head small; palpi porrect, extending about the length of head, and fringed with hair below; antennae of ♀ bipectinate, serrate at apex; frons with trilobate process from middle, the medial lobe longest; eyes large, rounded. Thorax clothed with hair; tibiae with tufts of hair. Abdomen with dorsal crest at base only. Fore wings with the apex rounded, the termen evenly rounded. Nervation: in fore wings, vein 2 given off slightly before two-thirds of cell; 3, 4 and 5 well separated, equidistant; discocellulars incurved; 6 from the lower cell-angle; areole narrow, elongate; 7 and 8 + 9 from the extremity thereof, 10 from its upper angle. In hind wings, vein 2 arising from five-sevenths of cell; 3, 4 and 6, 7 from a point at the angles of cell; discocellulars incurved; 5 obsolete; 8 anastomosing with the cell near base only.

Distribution: North America.

Eupseudomorpha brillians NEUMOEGEN

1880 *Edwardsia brillians* NEUMOEGEN, Canad. Entomologist, **12**:68. – Terra typica: Southwest Texas (Type, ♀, in Brooklyn Inst. Arts and Sci.).
1893 *Eupseudomorpha brillians* – DYAR, Canad. Entomologist, **25**:28.

Diagnosis: Head black, the palpi at base and lines round the eyes white; thorax black with dorsal and subdorsal white streaks; femora with tufts of orange hair and orange bands; tibiae with some orange scales; abdomen with the dorsal crest and anal tuft dark orange. Fore wings black; a short white fascia below costa towards base, a patch below middle of cell, a streak above medial part of dorsum, and a wedge-shaped patch below basal half of cell; a semicircular orange spot in middle of cell and a bar-shaped mark on discocellulars; its extremities rather dilated on outer side, with a triangular white patch between them; a white band beyond the cell from below costa to vein 2, its inner edge sinuous; a white postmedial band excurved from below costa to vein 2 where it is angled inwards, ending at vein 1; a subterminal fulvous-orange band with whitish outer edge from below costa to vein 1, expanding on inner side below vein 2; cilia white at extremities. Hind wings fuscous-black with a bronze tinge; faint traces of a discoidal orange spot and subterminal band. Length of fore wing 17 mm.

Distribution: U.S.A. (Texas).

Remark. – Following the communication kindly supplied by the British Museum (Natural History), the Noctuid genus *Eupseudomorpha* has been transferred by E. L. TODD to the Agaristidae. It would seem that this change has not yet been published (autumn 1974). For this reason, and also owing to the lack of information about the species, the present author places the genus *Eupseudomorpha* with the genera *incertae sedis.*

Note: *Vespola* WALKER, 1868 (Type species *Vespola caeruleifera* WALKER) does in the author's opinion not belong to the Agaristidae, and should be placed with the sister-family Noctuidae.

Alphabetic Register

acroleuca, Melanchroiopsis 67
adrasta, Diamuna 73
aedessa, Diamuna 74
aelia, Diamuna 74
–, Gerra 74
aeruginosa, Melanochroia 6
affinis, Erocha 15
aguirrei, Mystrocephala 7
–, Rhosus 7
albiceps, Mystrocephala 6
–, Rhosus 6
albidisca, Tuerta sevorsa 64
albifascia, Seirocastnia 32
albiplaga, Tuerta sevorsa 64
albirubra, Aucula hilzingeri 19
albomaculata, Noctua 49
Alypia 48
Alypiodes 56
Alypioides 56
amalthea, Aegocera 29
–, Othria 29
–, Phasis 29
–, Seirocastnia 29
amazonica, Othria 27
Androloma 48
argentata, Euthisanotia 43
–, Laquea 43
arguirrei, Alypia 7
Arpia 17
assimilis, Eudryas 76
Aucula 17
augias, Orthia 27
australis, Alypia 53

bartholomaei, Eudryas 70
batesii, Agarista 68
–, Eusemia 68
–, Phasis 68
Bepara 18
bimaculata, Agarista 57
–, Alypioides 57
–, dugesi, Alypiodes 57
–, Euchromia 50
–, Zygaena 49
bolicar, Neotuerta platensis 66
brannani, Alypia 56
–, Androloma 56
brephos, Gerra 74
brevipennis, Eudryas 80
–, Euthisanotia 80
brillians, Edwardsia 82
–, Eupseudomorpha 82
buprasia, Aucula 19
buprasium, Leiosoma 19

Caularis 69
Ciris 24
Cisaucula 9
clavigera, Melanochroia 5
Clitis 35
collectiora, Neotuerta sabulosa 65
colombiana, Mystrocephala 7
–, Rhosus 7
colombianus, Rhosus 7
columbina, Othria 32
–, Phasis 32
–, Seirocastnia 32
conjuncta, Alypia similis 55
continua, Josia 30
contraria, Phasidia 62
–, Phasis 62
Copidryas 40
cosyra, Copidryas 42
–, Euthisanotia 42
–, Misa 42
crescens, Alypoides 57
–, Eusemia 57
–, Pseudalypia 57
crotchii, Pseudalypia 81
cypris, Euthisanotia 76

Darceta 1, 35
denieri, Mystrocephala 8
–, Rhosus 8
desparata, Androloma 56
Diamuna 35
dipsaci, Agarista 52
–, Alypia 52
discifera, Euschirropterus 46
disciferus, Eucirrhopterus 46

discreta, Dioptis 12
disparata, Alypia 56
–, Androloma 56
disparilis, Heterandra 43
dolens, Erocha 16
–, Heterocampa 16
dugesi, Alypiodes bimaculata 57

ecuadoriana, Othria 3
–, Parothria 3
edwardsi, Agarista 55
Edwardsia 82
elaeina, Erocha 15
elaphebolia, Orthia 31
epimensis, Noctua 50
–, Psychomorpha 60
Epithisanotia 32
Erocha 1, 11
Eudryas 77
Eupseudomorpha 1, 82
euryrhoda, Psychomorpha 61
Euschirropterus 42
Euscirrhopterus 42
Euthisanotia 77
extensa, Seirocastnia 32

falcata, Darceta 38
–, Diamuna 38
–, Hecytesia 38
Fenaria 72
flavilinguis, Alypioides 57
furcifer, Schalifrontia 47
fuscus, Euschirropterus klagesi 45

geronimo, Alypioides 59
–, Pseudalypia 59
Gerra 72
Gerrodes 22
glaucopsis, Psychomorpha 50
gloveri, Copidryas 41
–, Euschirropterus 41
gracilenta, Alypia 56
grandimacula, Darceta 37
–, Diamuna 37
grata, Bombyx 76
–, Eudryas 76
–, Parathisanotia 76
grotei, Agarista 57
–, Alypia 57

haenschi, Darceta 39
hemicycla, Neotuerta 66
–, Tuerta 66
hesperica, Darceta 36
hesperina, Darceta 36
–, Diamuna 36
Heterandra 42
hilzingeri albirubra, Aucula 19
–, Aucula 19
–, Metagarista 19

hudsonica, Alypia 52
insulica, Tuerta 64
irrorata, Erocha 16
isabella, Rhosus 9

jamaicensis, Caularis 71
janeira, Arpia 20
–, Aucula 20
josioides, Aucula 18
juanita, Leiosoma 14

klagesi, Euschirropterus 45
–, Euscirrhopterus 45
–, fuscus, Euschirropterus 45
–, tucumanus, Euscirrhopterus 46

langtoni, Alypia 52
Laquea 42
Leiosoma 11
Leonides 11
Letaba 81
lethe, Agarista 27
–, Eusemia 27
leucodisca, Erocha 13
leuconoë, Mystrocephala 6
–, Pycnodontis 6
–, Rhosus 6
leucotelus, Erocha 13, 14
–, Euthisanotia 13
lindingii, Agarista 29
–, Eusemia 29
–, Othria 29
–, Phasis 29
longipes, Diamuna 22
–, Gerrodes 22
lorquini, Agarista 55
–, Alypia 55
–, Alypia maccullochii 55
–, Androloma 55
lulesa, Alypia 54
lunata, Alypia 53
–, Caularis 71
–, Gerra 75
luteomacula, Gerra sevorsa 74
lycaon, Neotuerta 66
–, Euthisanotia 66
–, Paraegocera 66
–, Tuerta 66

maccullochii, Alypia 54
– maccullochii, Alypia 55
–, Alypia maccullochii 55
– lorquini, Alypia 55
– similis, Alypia 55
magnifica, Aucula 20
–, Euthisanotia 20
mardava, Melanchroiopsis 69
–, Phasis 69
marginalis, Copidryas 20

mariposa, Agarista 53
–, Alypia 53
matuta, Alypia 51
Melanchroiopsis 67
meridiana, Phasis 31
–, Seirocastnia 31
meridionalis, Gerra sevorsa 74
–, Othria 62
mineata, Gerrodes 24
mummia, Bombyx 12
–, Eurocha 12
–, Pycnodontis 12
Mystrocephala 4

Neotuerta 63
nervalis, Seirocastnia praefecta 31
nexa, Orthia 27
–, Orthria 27
noctuiformis, Agarista 64
–, Tuerta 64
noa, Letaba 81
noctilux, Josia 68
–, Melanchroiopsis 68
–, Phasis 68
nocturna, Seirocastnia 31

octomaculata, Agarista 50
–, Alypia 49
–, Sesia 49
–, Sphinx 50
ophideres, Darceta 37
ornata, Alypia 54
Orthia 26
Othria 26
ovata, Mystrocephala 8
–, Pycnodontis 8
–, Rhosus 8
ovatus, Rhosus 8

panamensis, Seirocastnia 29
pampeana, Alypia 54
Parathisanotia 76
Parothria 2
peruviana, Cisaucula 10
–, Copidryas 10
Phasidia 61
Phasis 67
picta, Erocha 15
platensis bolivar, Neotuerta 66
–, Copidryas 65
–, Euthisanotia 65
–, Neotuerta 65
–, Neotuerta platensis 65
– platensis, Neotuerta 65
–, Tuerta 65
poeyi, Euschirropterus 43
–, Euscirrhopterus 43
– pulverosa, Euschirropterus 45
postica, Mystrocephala 6
posticus, Rhosus 5, 7

praefecta nervalis, Seirocastnia 31
–, Phasis 31
–, Seirocastnia 31
primulina, Clitis 37
–, Darceta 37
–, Diamuna 37
proserpina, Clitis 36
–, Darceta 36
–, Diamuna 36
–, Phalaena 36
Pseudalypia 1, 56, 80
Psychomorpha 59
pulchra, Gerra 75
pulverosa, Euschirropterus poeyi 5
–, Mystrocephala 7
–, Pycnodontis 7
pulverosus, Rhosus 7
Pycnodontis 4

quadriguttalis, Alypia 50
quadrimacula, Mystrocephala 6

radians, Agarista 58
–, Alypioides 58
–, Eusemia 58
–, Phalaenoides 58
radicalis, Gerra 73
Rhosus 4
ridingsi, Alypia 52
–, Androloma 52

sabulosa, Agarista 63
– collectiora, Neotuerta 65
–, Copidryas 64
–, Eusemia 63
–, Metagarista 63
–, Misa 64
–, Neotuerta 63
–, Neotuerta sabulosa 65
– sabulosa, Neotuerta 65
– Seudyra 64
–, Tuerta 64
sacramenti, Alypia 52
sanctaejohannis, Epithisanotia 33
–, Eudryas 33
sanctijohannis, Epithisanotia 33
–, Eudryas 33
Schalifrontia 47
schausi, Aucula 21
–, Eusemia 30
–, Phasis 30
Seirocastnia 28
semiviridis, Erocha 14
–, Euthisanotia 14
separata, Josia 29
–, Phasis 29
serpentium, Leiosoma 13
–, Leonides 14
severa, Darceta 39
–, Diamuna 39

–, Phalaena 39
sevorsa albidisca, Tuerta 64
– albiplaga, Tuerta 64
–, Fenaria 64, 74
–, Gerra 74
– luteomacula, Gerra 74
– meridionalis, Gerra 74
–, Phaegorista 74
similis, Alypia 55
–, Alypia maccullochii 55
–, Androloma 55
– conjuncta, Alypia 55
sophocles, Gerra 75
spadicea, Alypia 8
–, Mystrocephala 8
–, Pycnodontis 8
–, Rhosus 8
spadiceus, Rhosus 8
storniana, Mystrocephala 8
stornianus, Rhosus 8
stuartii, Pseudalypia 59
sublata, Aucula 21
–, Bepara 21

tribuna, Ephialtias 30
–, Phasis 30
–, Seirocastnia 30
trimaculata, Alypia 57
trita, Erocha 17
tucumanus, Euscirrhopterus klagesi 46
–, Euscirrhopterus walkeri 46
Tuerta 63

undulans, Caularis 70, 71
unio, Eudryas 78
–, Euthisanotia 78
unipuncta, Rhosus 8

Vespola 82
volupia, Othria 30
–, Seirocastnia 30

walkeri, Alypioides 59
–, Euschirropterus 45
–, Euscirrhopterus 45
–, Euscirrhopterus walkeri 46
–, Pseudalypia 59
– tucumanus, Euscirrhopterus 46
– walkeri, Euscirrhopterus 46
wilsonii, Ciris 25
–, Xerociris 25
wittfeldi, Alypia 51

Xerociris 24

DAS TIERREICH

Eine Zusammenstellung und Kennzeichnung der rezenten Tierformen

Herausgeber
Robert Mertens und Willi Hennig
Schriftleiter: Heinz Wermuth

Lfg.	Bearbeiter	Titel	Ersch.-jahr	Umfang	Preis in DM
45	A. Schmidt	COLEOPTERA: *Aphodiinae*	1922	VI, 614 Seiten, 34 Abbildungen	120,–
46	F. Nieden	AMPHIBIA: *Anura I*	1923	XXXII, 584 Seiten, 380 Abbildungen	115,50
47	W. Kükenthal	COELENTERATA: *Gorgonaria*	1924	XXVIII, 478 Seiten, 209 Abbildungen	144,–
48	J. J. Kieffer	HYMENOPTERA: Proctotrupoidea: *Scelionidae*	1926	XXXVI, 885 Seiten, 340 Abbildungen	300,–
49	F. Nieden	AMPHIBIA: *Anura II Engystomatidae*	1926	XVI, 119 Seiten, 55 Abbildungen	42,–
50	E. Giglio-Tos	ORTHOPTERA: *Mantidae*	1927	XL, 707 Seiten, 6 Abbildungen	247,50
51	H. Stichel	LEPIDOPTERA: *Nemeobiinae*	1928	XXX, 300 Seiten, 197 Abbildungen	126,50
52	C. Graf Attems	MYRIAPODA: *I. Geophilomorpha*	1929	XXIII, 388 Seiten, 307 Abbildungen	180,–
53	F. Kiefer	CRUSTACEA COPEPODA: *II. Cyclopoida Gnathostoma*	1929	XVI, 102 Seiten, 42 Abbildungen	56,30
54	C. Graf Attems	MYRIAPODA: *II. Scolopendromorpha*	1930	XIX, 308 Seiten, 387 Abbildungen	150,–
55	E. Ahl	AMPHIBIA: *Anura III Polypedatidae*	1931	XVI, 477 Seiten, 320 Abbildungen	232,50
56	S. Thor	ARACHNOIDEA ACARINA: *Bdellidae, Nicoletiellidae, Cryptognathidae*	1931	XIII, 87 Seiten, 93 Abbildungen	56,20
57	M. Beier	PSEUDOSCORPIONIDEA I: Subord: *Chthoniinea* et *Neobisiinea*	1932	XX, 258 Seiten, 271 Abbildungen	120,–
58	M. Beier	PSEUDOSCORPIONIDEA II: Subord. *Cheliferinea*	1932	XXI, 294 Seiten, 300 Abbildungen	146,20

Lfg.	Bearbeiter	Titel	Ersch.-jahr	Umfang	Preis in DM
59	M. Holly	CYCLOSTOMATA	1933	XIV, 62 Seiten, 57 Abbildungen	37,5[illegible]
60	S. Thor	ACARINA: *Tydeidae, Ereynetidae*	1933	XI, 84 Seiten, 102 Abbildungen	45,–
61	M. Holly	PISCES: *2. Dipnoi*	1933	IX, 20 Seiten, 25 Abbildungen	10,–
62	F. Werner	REPTILIA: *Loricata*	1933	XIV, 40 Seiten, 33 Abbildungen	26,2[illegible]
63	M. Holly	PISCES: *4. Crossopterygii*	1933	X, 23 Seiten, 30 Abbildungen	10,–
64	F. Bryk	LEPIDOPTERA: *Baroniidae, Teinopalpidae, Parnassiidae I*	1934	XXIII, 131 Seiten, 87 Abbildungen	75,–
65	F. Bryk	LEPIDOPTERA: *Parnassiidae II* (Subfam. *Parnassiinae*)	1935	LI, 790 Seiten, 698 Abbildungen	412,50
66	E. Marcus	ARTHROPODA: *Tardigrada*	1936	XVI, 329 Seiten, 306 Abbildungen	150,–
67	M. Holly	PISCES: *4. Ganoidei*	1936	XVI, 65 Seiten, 56 Abbildungen	37,50
68	C. Graf Attems	MYRIAPODA: *3. Polydesmoidea I.* Fam. *Strongylosomidae*	1937	XXII, 300 Seiten, 343 Abbildungen	150,–
69	C. Graf Attems	MYRIAPODA: *3. Polydesmoidea II.* Fam. *Leptodesmidae, Platyrhachidae, Oxydesmidae, Gomphodesmidae*	1938 1938	XXVIII, 487 Seiten, 509 Abbildungen	243,80
70	C. Graf Attems	MYRIAPODA: *3. Polydesmoidea III*	1940	XXXII, 568 Seiten, 719 Abbildungen	300,–
71	S. Thor & C. Willmann	ACARINA: *3.–71a: Eupodidae, Penthalodidae, Penthaleidae, Rhagidiidae, Pachygnathidae, Cunaxidae. – 72b Trombidiidae*	1941/1947	XXXVI, 541 Seiten, 599 Abbildungen	389,20
72	H. Wermuth	REPTILIA: *Helodermatidae*	1958	16 Seiten, 9 Figuren	16,–
73	M. Beier	ORTHOPTERA: *Tettigoniidae* (Pseudophyllinae I)	1962	XII, 468 Seiten, 245 Abbildungen	210,–
74	M. Beier	ORTHOPTERA: *Tettigoniidae* (Pseudophyllinae II)	1960	396 Seiten, 241 Figuren	262,50
75	K. Odening	TREMATODA (Digenea): *Plagiorchiidae* III (Haematoloechinae) und *Omphalometridae*	1960	IV, 77 Seiten, 60 Figuren	64,–

-fg.	Bearbeiter	Titel	Ersch.-jahr	Umfang	Preis in DM
76	K. Sanft	AVES (Upupae): *Bucerotidae*	1960	IV, 176 Seiten, 106 Figuren	156,–
77	R. Kilias	GASTROPODA PROSOBRANCHIA: *Tonnidae*	1962	63 Seiten, 47 Figuren	28,–
78	St. W. Gorham	GYMNOPHIONA. *Liste der rezenten Amphibien und Reptilien*	1962	X, 25 Seiten	15,–
79	R. Mertens	HELODERMATIDAE, VARANIDAE, LANTHANOTIDAE. *Liste der rezenten Amphibien u. Reptilien*	1963	X, 26 Seiten	14,–
80	H. Wermuth	GEKKONIDAE, PYGOPODIDAE, XANTUSIIDAE. *Liste der rezenten Amphibien und Reptilien*	1965	XXII, 246 Seiten	117,80
81	J. A. Peters	COLUBRIDAE (Dipsadinae). *Liste der rezenten Amphibien und Reptilien*	1965	VIII, 18 Seiten	12,50
82	J. Illies	PLECOPTERA, *Katalog*	1966	XXX, 632 Seiten, 20 Abbildungen	294,–
83	R. Mertens	CHAMAELEONIDAE. *Liste der rezenten Amphibien und Reptilien*	1966	X, 37 Seiten	26,–
84	C. Gans	UROPELTIDAE. *Liste der rezenten Amphibien und Reptilien*	1966	29 Seiten	14,–
85	St. W. Gorham	ASCAPHIDAE, LEIOPELMATIDAE, PIPIDAE, DISCOGLOSSIDAE, PELOBATIDAE, LEPTODACTYLIDAE, RHINOPHRYNIDAE. *Liste der rezenten Amphibien und Reptilien*	1966	XVI, 222 Seiten	107,–
86	H. Wermuth	AGAMIDAE. *Liste der rezenten Amphibien und Reptilien*	1967	XIV, 127 Seiten	72,–
87	H. Wermuth	CORDYLIDAE (Cordylinae + Gerrhosaurinae). *Liste der rezenten Amphibien und Reptilien*	1968	X, 30 Seiten	24,–
88	F. Haas	BIVALVIA: *Unionacea*	1969	X, 663 Seiten, 5 Abbildungen	380,–
89	A. F. Stimson	BOIDAE, (Boinae + Bolyeriinae + Loxoceminae + Pythoninae). *Liste der rezenten Amphibien und Reptilien*	1969	XI, 49 Seiten	38,–

Lfg.	Bearbeiter	Titel	Ersch.-jahr	Umfang	Preis in DM
90	H. Wermuth	ANGUIDAE, ANNIELLIDAE, XENOSAURIDAE. *Liste der rezenten Amphibien und Reptilien*	1969	XII, 41 Seiten	33,–
91	M. Fischer	HYMENOPTERA: *Braconidae* (Opiinae I)	1972	XII, 620 Seiten, 463 Figuren	580,–
92	R. Kilias	GASTROPODA PROSOBRANCHIA: *Cymatiidae*	1973	VIII, 235 Seiten, 149 Figuren	288,–
93	P. Banarescu & T. T. Nalbant	PISCES, TELEOSTEI: *Cyprinidae (Gobioninae)*	1973	VII, 304 Seiten, 154 Figuren, 19 Karten	320,–
94	P. Zwick	PLECOPTERA, *Phylogenetisches System und Katalog*	1973	XXXII, 465 Seiten, 75 Abbildungen	490,–
95	E. Duellman	HYLIDAE, CENTROLENIDAE, PSEUDIDAE. *Liste der rezenten Amphibien und Reptilien*	1976	XIX, 225 Seiten	280,–
96	M. Fischer	HYMENOPTERA: *Braconidae* (Opiinae II / Amerika)	1976	etwa 1038 Seiten, 890 Figuren	980,–
97	S. G. Kiriakoff	LEPIDOPTERA NOCTUIFORMES: *Agaristidae I*	1976	IX, 180 Seiten, 39 Figuren	150,–
98	S. G. Kiriakoff	LEPIDOPTERA NOCTUIFORMES: *Agaristidae II*	1976	VIII, 165 Seiten, 52 Figuren	140,–
99	S. G. Kiriakoff	LEPIDOPTERA NOCTUIFORMES: *Agaristidae III*	1976	VI, 86 Seiten, 26 Figuren	90,–
100	H. Wermuth & † R. Mertens	TESTUDINES, CROCODYLIA, RHYNCHOCEPHALIA. *Liste der rezenten Amphibien und Reptilien*	1976	XXVII, 174 Seiten	210,–

FSC
www.fsc.org
MIX
Papier | Fördert
gute Waldnutzung
FSC® C083411

Zeitfracht Medien GmbH
Ferdinand-Jühlke-Straße 7
99095 Erfurt, Deutschland
produktsicherheit@kolibri360.de